Computational Models of Motivation
for Game-Playing Agents

Kathryn E. Merrick

Computational Models of Motivation for Game-Playing Agents

 Springer

Kathryn E. Merrick
School of Engineering and Information
   Technology
University of New South Wales
   Australian Defence Force Academy
Canberra, ACT
Australia

ISBN 978-3-319-81518-3   ISBN 978-3-319-33459-2   (eBook)
DOI 10.1007/978-3-319-33459-2

Printed on acid-free paper

This Springer imprint is published by Springer Nature
The registered company is Springer International Publishing AG Switzerland

Preface

In recent years, two quests have been emerging to focus artificial intelligence research for games. The first, and newest, is concerned with developing our understanding of game players. The second is concerned with building more believable and exciting virtual worlds in which to play games. In this book we use information gathered by game data mining researchers about players to inform the design of novel self-motivated game-playing agents to control non-player characters. We demonstrate how self-motivated agents can increase the diversity of non-player characters by permitting agents to exhibit unique decision-making characteristics, which lead to interesting, emergent patterns when they interact.

More and more studies are emerging of the factors that motivate people to play online games and the cultures that emerge among humans in virtual worlds. However, the diversity we see in humans is not yet present in the computer-controlled characters that support online virtual worlds. Techniques for embedding human-like motives in game-playing agents have not yet been widely explored. As the complexity and functionality of multiuser virtual worlds increases, computer-controlled characters are becoming an increasingly challenging application for artificial intelligence techniques. Players are demanding more believable and intelligent non-player characters to enhance their gaming experience.

This book presents a new artificial intelligence technique—computational motivation—and shows how it can be used to increase the diversity of non-player characters by adapting some traditional architectures for computer-controlled game characters. Theoretical issues are addressed for representing and embedding computational models of motivation in rule-based agents, crowds, learning agents and evolutionary algorithms. Practical issues are addressed for defining games, mini-games or in-game scenarios for virtual worlds in which computer-controlled, motivated agents can participate alongside human players.

Large-scale virtual worlds may have hundreds or thousands of virtual characters, and the question arises as to how these characters can be embedded with diverse, individual personalities. Computational motivation provides a novel answer to this question. Our starting point is the 'influential trio': achievement, affiliation and power motivation. Incentive-based theories of achievement, affiliation and power motivation are the basis of competence-seeking behaviour, relationship-building, leadership and resource-controlling behaviour in humans. We show how these motives can be modelled and embedded in artificial agents.

The aim of this book is to provide game programmers, and those with an interest in artificial intelligence, with the knowledge required to develop diverse, believable game-playing agents for virtual worlds. Computational motivation is an exciting, emerging research topic in the field of artificial intelligence. The development of motivated agents is at the cutting edge of artificial intelligence and cognitive modelling research. This opens the way both for new types of artificial agents, new types of computer games and in-world mini-games or scenarios. This book provides an in-depth look at new computational models of motivation and offers insights into the strengths, limitations and future development of motivated agents for gaming applications.

Part I—Game Playing in Virtual Worlds by Humans and Agents

Chapter 1—*From Player Types to Motivation*
Chapter 2—*Computational Models of Achievement, Affiliation and Power Motivation*
Chapter 3—*Game-Playing Agents and Non-Player Characters*

The first part of this book studies the relationship between game play and motivation in humans and proposes ways in which this can be represented and embedded in artificial agents. Chapter 1 studies the motivational characteristics we see in humans playing games, which we might expect to see in a diverse society of computer-controlled game-playing agents. After examining player types that have been identified through subjective and objective studies of human game players, Chap. 1 turns to complementary literature from motivation psychology and reviews the theories that may contribute to these characteristics in humans. It specifically focuses on three incentive-based theories of motivation for achievement, affiliation and power motivation.

Chapter 2 introduces computational models of motivation to embed these human-inspired motives in artificial agents at different levels of fidelity. A flexible mathematical model is introduced that permits achievement, affiliation and power motives to be expressed in terms of approach and avoidance components, which can be adjusted to create different motivation variants. Alternatives are provided to model individual motives, or profiles of several motives.

Chapter 3 describes how motivation can be embedded in four architectures for game-playing agents that can be used to control non-player characters: rule-based agents, crowds, learning agents and evolutionary algorithms. Motivated rule-based agents are suitable for decision making by individual agents, while motivated learning agents are suitable for competitive or strategic decision making when two or more agents interact. The motivated crowd and evolutionary algorithms are suitable for controlling groups of agents. These architectures are the topics of study in Part II, Part III and Part IV of the book.

Part II—Comparing Human and Artificial Motives

Chapter 4—*Achievement Motivation*
Chapter 5—*Profiles of Achievement, Affiliation and Power Motivation*

Part II of this book describes how the models introduced in Part I can be used to create a number of specific motivation subtypes that have previously been observed in humans. These are then demonstrated by reproducing three canonical human experiments with artificial agents. In Chap. 4, four subtypes of achievement motivation are described. These are embedded in agents playing the ring-toss game. We demonstrate the similarities that can be observed between motivated game-playing agents and humans playing the same game.

In Chap. 5 three profiles of achievement, affiliation and power motivation are introduced for agents playing two different games: roulette and the prisoners' dilemma game. We again demonstrate the similarities that can be observed between motivated game-playing agents and humans playing the same games. We show that we can create a diverse group of agents that compete or cooperate in different ways when playing games.

Part III—Game Scenarios for Motivated Agents

Chapter 6—*Enemies*
Chapter 7—*Pets and Partner Characters*
Chapter 8—*Support Characters*

In Part III in-game scenarios are described that are appropriate for different kinds of motivated agents, including motivated rule-based agents, crowds and learning agents. These scenarios are presented in three chapters, corresponding to scenarios for three common non-player character types: enemies, partner characters and support characters.

Chapter 6 presents an application of motivated rule-based agents as enemies and two scenarios in which motivated learning agents are used as enemies. Chapter 7 considers two further scenarios for motivated learning agents as pets and partner characters. Chapter 8 explores the use of motivated crowds and motivated learning agents in support characters. We show that agents with different motives exhibit different strategies for playing games. This results in behavioural diversity among characters. In addition, when agents with different motives interact, different outcomes emerge.

Included in this part are theoretical analyses, empirical investigations and sample game applications. We show how theoretical and empirical results can be used to predict the behaviour of agents in practice. We use these motivated agents in two ways: first as non-player characters and secondly to model the different types of responses possible to non-player characters, either by individuals or groups of human players. The applications in this part of the book demonstrate how each abstract agent architecture from the previous part can be connected to the concrete actions available to a game character.

Part IV—Evolution and the Future of Motivated Agents

Chapter 9—*Evolution of Motivated Agents*
Chapter 10—*Conclusion and Future*

Part IV looks to the future of motivated game-playing agents, first with a specific focus on the conditions under which agents with different motives might evolve in a virtual world and then more broadly. Chapter 9 considers the evolution of motivation in a society of game-playing agents. Agents are studied in multiplayer social dilemma games and in a custom implementation of a classic single-player shooter game. This chapter demonstrates how the composition of a society of motivated agents can change over time in response to subjective or objective definitions of the fitness of an agent.

Chapter 10 summarises the algorithms, components and combinations of components that have been studied in the book, and discusses how these components might fit into other motivated agent architectures. The strengths and limitations of motivated agents are considered and used as a basis for discussion of the future directions for motivated agents and multiuser computer games. Advances in computational models of motivation, motivated agent models and their application to different types of games are considered.

Canberra, ACT, Australia   Kathryn E. Merrick

Contents

Part I Game Playing in Virtual Worlds by Humans and Agents

1 From Player Types to Motivation 3
  1.1   Virtual Worlds and Online Games 3
  1.2   Explorer, Achiever, Socialiser, Aggressor: Human Player
       Types 4
  1.3   Incentive-Based Theories of Motivation 9
       1.3.1   Incentive 9
       1.3.2   Achievement Motivation 10
       1.3.3   Affiliation Motivation 13
       1.3.4   Power Motivation 14
       1.3.5   Dominant Motives and Motive Profiles 15
       1.3.6   Motivation and Zeitgeist 16
  1.4   Conclusion 17
  References... 18

**2 Computational Models of Achievement, Affiliation
and Power Motivation**...................................... 21
  2.1   Towards Computational Motivation 21
  2.2   Modelling Incentive-Based Motives Using
       Approach-Avoidance Theory 22
       2.2.1   Modelling Achievement Motivation 26
       2.2.2   Modelling Affiliation Motivation.................. 31
       2.2.3   Modelling Power Motivation..................... 33
  2.3   Motive Profiles for Artificial Agents...................... 34
       2.3.1   Modelling Profiles of Achievement, Affiliation
            and Power....................................... 35
       2.3.2   Modelling the Dominant Motive Only.............. 36
       2.3.3   Optimally Motivating Incentive................... 36
  2.4   Using Motive Profiles for Goal Selection 39
       2.4.1   Winner-Takes-All 40
       2.4.2   Probabilistic Goal Selection...................... 41

2.5   Summary   41
References.   42

3   **Game-Playing Agents and Non-player Characters**   45
3.1   Artificial Intelligence in Non-player Characters   45
3.2   Rule-Based Agents.   46
3.2.1   Game Scenarios for Motivated Rule-Based Agents   47
3.2.2   Motivated Rule-Based Agents.   48
3.3   Crowds.   49
3.3.1   Motivated Crowds.   51
3.4   Learning Agents.   52
3.4.1   Social Dilemma Games.   53
3.4.2   Strategies in Game Theory   55
3.4.3   Motivated Learning Agents.   56
3.5   Evolution   59
3.5.1   Multiplayer Social Dilemma Games   59
3.5.2   Evolution in Multiplayer Social Dilemma Games   60
3.5.3   Evolution of Motivated Agents   60
3.6   Summary   64
References.   64

Part II   **Comparing Human and Artificial Motives**

4   **Achievement Motivation**   69
4.1   Scenarios and Mini-games for Motivated Agents   69
4.2   The Ring-Toss Game   70
4.3   Modelling Achievement-Motivated Rule-Based Agents   72
4.4   AchAgents Playing the Ring-Toss Game   75
4.4.1   Comparison to Humans.   75
4.4.2   Comparison to Atkinson's Risk-Taking Model   77
4.4.3   Comparison to a Model of Achievement Motivation
with Learning   79
4.5   Conclusion   80
References.   82

5   **Profiles of Achievement, Affiliation and Power Motivation**   83
5.1   Evolving Motivated Agents to Play Multiple Games   83
5.2   Roulette   84
5.2.1   Motive Profiles for Roulette   85
5.2.2   Motivated Rule-Based Agents Playing Roulette   87
5.2.3   Comparison to Humans.   89
5.3   The Prisoners' Dilemma Game.   90
5.3.1   Motivated Rule-Based Agents and the Prisoners'
Dilemma   92
5.3.2   Comparison to Humans.   92
5.4   Conclusion   94
References.   95

Part III   Game Scenarios for Motivated Agents

6   Enemies   99
   6.1   Types of Non-player Characters and Their Roles   99
   6.2   *Paratrooper*   100
       6.2.1   Motivated Rule-Based Agents as Paratroopers.   101
       6.2.2   Empirical Study: Motivated Paratroopers.   103
       6.2.3   Summary.   104
   6.3   The Prisoners' Dilemma Revisited   105
       6.3.1   Arms Races in Turn-Based Strategy Games   106
       6.3.2   Perception of Arms Races by Motivated Learning
             Agents.   106
       6.3.3   Empirical Study: Motivated Learning Agents in an
             Arms Race.   110
   6.4   The Leader Game.   113
       6.4.1   Perception of Leader Games by Motivated Learning
             Agents.   115
       6.4.2   Empirical Study: Motivated Learning Agents as
             Explorers.   117
       6.4.3   Application: Motivation to Explore in a Turn-Based
             Strategy Game.   119
   6.5   Conclusion   124
   References.   125

7   Pets and Partner Characters   127
   7.1   Pets, Partners and Minions   127
   7.2   The Chicken (or Snowdrift) Game   128
       7.2.1   Perception of the Snowdrift Game by Motivated
             Learning Agents   129
       7.2.2   Empirical Study: Motivated Learning Agents
             as Partner Characters.   131
       7.2.3   Application: Motivated Battle Pets or Minions   133
   7.3   Battle of the Sexes   137
       7.3.1   Perception of Battle of the Sexes by Motivated
             Learning Agents   138
       7.3.2   Empirical Study: Motivated Learning Agents
             as Negotiators   140
   7.4   Conclusion   142
   References.   143

8   Support Characters   145
   8.1   Motivation in Multi-agent Systems.   145
   8.2   *Breadcrumbs*.   146
       8.2.1   Attracting a (Motivated) Crowd.   146
       8.2.2   Case Studies   147
       8.2.3   Scenarios Abstracted by Breadcrumbs.   150

8.3   Chicken Revisited.   151
        8.3.1   Player-Base Motive Indices.   151
        8.3.2   Motivated Learning Agents as Vendors.   152
        8.3.3   Empirical Study: Vendors Bargaining
                  with Multiple Opponents.   153
8.4   Battle of the Sexes Revisited   155
        8.4.1   Motivated Learning Agents as Quest Givers   155
        8.4.2   Empirical Study: Quest Givers Negotiating
                  with Multiple Opponents.   158
8.5   Conclusion   158
References.   159

Part IV   Evolution and the Future of Motivated Agents

9   Evolution of Motivated Agents.   163
9.1   The Evolutionary Perspective   163
9.2   Multiplayer Social Dilemma Games   164
        9.2.1   Common Pool Resource Games   168
        9.2.2   n-Player Leader   171
        9.2.3   The Hawk-Dove Game   173
        9.2.4   n-Player Battle of the Sexes   176
        9.2.5   Summary.   180
9.3   *Paratrooper* Revisited   180
        9.3.1   Evolution of Motivated Paratroopers   180
        9.3.2   Case Study   181
9.4   Conclusion   182
References.   183

10   Conclusion and Future   185
10.1   The Building Blocks of Computational Motivation   185
        10.1.1   Feature Selection.   186
        10.1.2   Task or Goal Generation   187
        10.1.3   Motivation Functions.   188
        10.1.4   Arbitration Functions.   189
        10.1.5   Goal Selection.   189
        10.1.6   Summary.   190
10.2   Motivated Agents   190
        10.2.1   Motivated Rule-Based Agents   190
        10.2.2   Motivated Learning Agents   191
        10.2.3   Motivated Reinforcement Learning Agents   191
        10.2.4   Crowds of Motivated Agents.   192
        10.2.5   Evolution of Motivated Agents   192
        10.2.6   Summary.   192

10.3   Future   193
        10.3.1   Motivated Agents in Computer Games   193
        10.3.2   The Future of Computational Motivation   194
    References.   195

Appendix A: Transformations of Social Dilemma Games   197

Appendix B: Flocking Parameter Values Used in *Breadcrumbs*   207

Index   209

Symbols and Acronyms

A^j	An agent (human or computer-controlled) with identifying label j
B^i	A behaviour (skill or sequence of actions) with identifying label i
C	A condition
CPR	Common-pool resource
$E^j(B)$	Expected payoff when agent A^j executes behaviour B
G^g	A goal with identifying label g
$f^k(x)$	Fitness of the kth type of agent
$F^k(B_t = B^i)$	Fraction of agents of type k executing behaviour B^i at time t
$I^f(G)$	Explicit incentive to avoid failure at goal G
$I^s(G)$	Explicit incentive for successfully achieving goal G
$\hat{I}^s(G)$	Subjective (perceived) incentive for successfully achieving goal G
MMORPG	Massively multiplayer, online role-playing game
MMOW	Massively multiplayer, online world
MUD	Multiuser dungeon
NPC	Non-player character
nPD	n-Player prisoners' dilemma
OMI	Optimally motivating incentive
Ω^j	Optimally motivating incentive of agent A^j
Ω^k	Optimally motivating incentive of agent type k
PD	Prisoners' dilemma
$P^j(B_t = B^i)$	Probability that the behaviour executed by agent A^j at time t is B^i
$P^j(G_t = G^g)$	Probability that the goal selected by agent A^j at time t is G^g
$P^s(G)$	Probability of successfully achieving goal G
RL	Reinforcement learning
RTM	Risk-taking model
t	Time

TBS	Turn-based strategy
T^{res}	Resultant tendency for motivation
V	Payoff value
W	An abstract game 'world' (mini-game or scenario)
$\widehat{\mathbf{W}}$	Perceived game world

List of Algorithms

Algorithm 3.1 A motivated rule-based agent with winner-takes-all
              goal selection 49
Algorithm 3.2 A motivated rule-based agent with probabilistic goal
              selection 49
Algorithm 3.3 A crowd of motivated agents.................... 52
Algorithm 3.4 A motivated learning agent 58
Algorithm 3.5 Evolving the proportions of agents with different
              motives when fitness is determined objectively 63
Algorithm 3.6 Evolving the proportions of agents with different
              motives when fitness is determined subjectively........ 63

Part I
Game Playing in Virtual Worlds by Humans and Agents

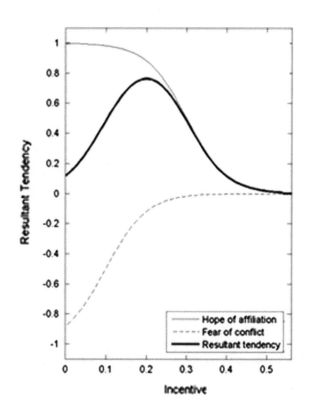

Chapter 1
From Player Types to Motivation

The first part of this book studies the relationship between game play and motivation in humans and proposes ways that motivation can be represented and embedded in artificial agents. This chapter studies the motivational characteristics we see in humans playing games, which we might expect to see in a diverse society of computer-controlled game-playing agents. After studying player types that have been identified through subjective and objective studies of human game players, the chapter turns to complementary literature from motivation psychology and reviews the theories that may contribute to these characteristics in humans. It specifically focuses on three incentive-based theories of motivation for achievement, affiliation and power motivation.

1.1 Virtual Worlds and Online Games

The term virtual world is now largely associated with computer-based, interactive, 3D simulated environments, where human users take the form of avatars. Virtual worlds may support as few as one human user playing a game, or as many as millions playing, socializing or even doing business [17, 34]. These users, or players, interact with each other, as well as with computer-controlled non-player characters. Non-player characters might take on roles as storytellers, teachers, merchants, friends, enemies, or all of the above [20], depending on the nature of the virtual world.

As the complexity and functionality of virtual worlds increase, the size and demographic diversity of virtual world users is also increasing [18]. There is a need to tailor virtual environments to support users with different skills and past experiences. This becomes particularly important when virtual worlds, previously used for games and social interaction, are adapted for education [5], training [21] and other 'serious' purposes. New techniques such as player experience modelling and game data mining are emerging as researchers seek to understand the motives of

© Springer International Publishing AG 2016

K.E. Merrick, *Computational Models of Motivation for Game-Playing Agents*,
DOI 10.1007/978-3-319-33459-2_1

virtual world users in general and online game players in particular [38]. Player experience modelling [33, 37], for example, uses computational techniques to construct models of players' experiences and satisfaction levels. Related to this, game data mining [12, 39] seeks a better understanding of how and why players play games.

Studies are increasingly emerging of the factors that motivate people to play online games [12, 39] and the cultures that emerge among humans in virtual worlds [9]. However, the diversity we see in human players is not yet present in the computer-controlled non-player characters that support online virtual worlds. Humans playing online games not only strive to master the game, but may also display personalised, idiosyncratic behaviours for collecting and stockpiling game resources, exploring the game world and building relationships with other players. Human players in the same situation act differently as a result of their motives and the preferences for different kinds of incentives that these motives inspire.

Techniques for embedding these kinds of motives in game-playing agents that control non-player characters have not yet been widely explored. To model these kinds of preferences for different behaviour in artificial agents, we first need to understand the different kinds of behaviour and the motives that cause them. We can then develop this understanding into computational models that can be embedded in game-playing agents that control non-player characters.

This chapter begins by examining the literature of game data mining for player types and player motivations that have been identified among those who play games in multiuser virtual worlds. Section 1.3 then refers to the literature of psychological motivation to examine theories that can support and explain these player types and motives. This provides the background necessary to develop computational models of motivation for game-playing agents in Chap. 2.

1.2 Explorer, Achiever, Socialiser, Aggressor: Human Player Types

One of the earliest and most well-known taxonomies of player types for online virtual worlds is Bartle's *Hearts, Clubs, Diamonds, Spades: Players Who Suit MUDs* [7]. A MUD is a multiuser dungeon, the precursor to contemporary massively multiuser online worlds (MMOWs) and massively multiplayer, online role-playing games (MMORPGs). Bartle's [7] taxonomy proposes four player types: achievers, explorers, socialisers and killers. These types were established by categorising players in two dimensions as shown in Fig. 1.1. The first dimension (on the horizontal axis) is a spectrum of interest that ranges from interest in other players to interest in the game world. It has some commonality with the axes of well-known personality type charts, which contrast terms such as 'people-oriented' against 'task-oriented' or 'personal' against 'logical' [10, 31].

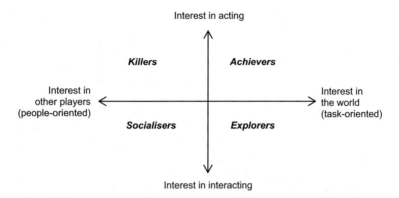

Fig. 1.1 Bartle [7, 8] uses an interest graph similar to this one to classify players as achievers, explorers, socialisers and killers. Classification depends on where players' interests lie on a players-world spectrum (*horizontal axis*) and an acting-interacting spectrum (*vertical axis*)

The second dimension (on the vertical axis) is a spectrum of interest that ranges from interest in acting on the game world or other players to interacting with the game world or other players. Achievers are defined as those who like 'acting on the world'. They like to master the game and achieve within the context defined by the game. Examples of such achievement include earning points or winning prizes.

Explorers, in contrast to achievers, like to 'interact with the world'. They like to gain a detailed knowledge of the world and how it works. Likewise, socialisers like to 'interact with other players'. For example they might make use of chat and emote functions provided in the game world. Earning points and winning prizes are less important to explorers and socialisers than to achievers.

Finally, killers are defined as those who are interested in doing things to people or 'acting on players'. As the name suggests, acting on players often takes the form of 'killing' other players' avatars, although it may also include gaining control of significant resources or winning an in-game competition or mini-game.

Bartle [7] further argued that it may be possible to influence the types of players who will favour a particular virtual world by changing the emphasis of the game world. For example, provision of multiple communication channels might favour those who have more interest in other players than in the game world. Making the world large and difficult to traverse might favour those who have more interest in the game world. Complex class systems and achievement systems in contemporary MMORPGs might favour individuals with more interest in acting than interacting. The provision of mini-games and puzzles is hypothesised to support the interaction end of the interest spectrum.

In addition to changes in the emphasis of the game world, changes in the underlying distribution of player types in the population might also influence the future composition of the population. Bartle [7] proposed an influence diagram to show how changes in the number of each player type might affect the number of players of other types. For example, he hypothesised that increasing the number of killers in the population will lead to a decrease in the number of achievers, explorers and socialisers (who don't like to be killed). Alternatively, an increase in the number of socialisers in the game might attract more socialisers (who like to talk to each other) and may decrease the number of killers. The assumption of Bartle's [7] model is that preference by a player for one type of play suppresses other types of play.

Bartle's [7] four player types emerged from a discussion by expert players in an early MUD. His model is thus an example of an early subjective player model. Analysis of player types and the way in which they are supported or discouraged by game design continues to be a relevant topic on game-related forums and blogs. A 2008 blog post on cooperation and competition in MMORPGs [15] differentiates between worlds with finite resources, which the author calls 'zero-sum games', and worlds of infinite resources, or 'infinite games'. In zero-sum or finite-resource worlds, winning is defined as accumulating more of some scarce resource than anyone else. The author argues that this supports only competitive play. In contrast, infinite worlds encourage players to add to the pool of resources and create new things. They can permit more than one 'winner'. The author goes on to argue that most contemporary games worlds tend to focus on controlling users' experiences rather than permitting them to create new content. The result is bias towards worlds that support competitive rather than cooperative (social) play.

More recent work is striving to link player types empirically with player demographics and build more objective models. Yee [39], for example, took a factor analytic approach to create an empirical model of player motivations. He surveyed 3,000 MMORPG players from games such as *EverQuest* [2], *Dark Age of Camelot* [35], *Ultima Online* [1] and *Star Wars Galaxies* [3]. His study revealed 10 motivation subcomponents that he grouped into three overarching components for achievement, socialising and immersion. Yee's grouping is shown in Table 1.1.

Table 1.1 The subcomponents of motivation identified by Yee [39]

Achievement	Social	Immersion
Advancement	*Socialising*	*Discovery*
Progress, power, accumulation, status	Casual chat, helping others, making friends	Exploration, lore, finding hidden things
Mechanics	*Relationship*	*Role-Playing*
Numbers, optimisation, templating, analysis	Personal, self-disclosure, find and give support	Story line, character history, roles, fantasy
Competition	*Teamwork*	*Customisation*
Challenging others, provocation, domination	Collaboration, groups, group achievements	Appearance, accessories, style, colour schemes
		Escapism
		Relax, escape from reallife, avoid real-life problems

Yee groups these subcomponents into three main components called achievement, social and immersion

The achievement component of motivation has three subcomponents. The first is a desire for advancement (progress) in the game and accumulation of in-game symbols of wealth or status. The second is an interest in analysing the underlying rules and game system to optimise performance in the game. The third is the desire to challenge and compete with others.

The socialising component of motivation, according to Yee's [39] study, also has three subcomponents: the desire to interact with and assist other players; the desire to form long-term and meaningful relationships with others; and the satisfaction derived from being part of a team while playing the game.

Finally, the immersion component has four subcomponents. The first is a desire to discover things that other players don't know about. The second is an interest in creating a persona and interacting with other players to create an improvised story. This includes customising the appearance of one's avatar, which is considered important enough to be a subcomponent in its own right. The last subcomponent of immersion is escapism—the desire to use the online environment to avoid thinking about real-life problems.

In another example, Drachen et al. [11] use unsupervised learning to derive player types from game-play features. Values for these features were collected from 1,365 players who completed *Tomb Raider: Underworld* [4], a single-player action-adventure game. Drachen et al. [11] identified four types of players they called veterans, solvers, pacifists and runners. Veterans are the most proficient group of players. Despite dying often, they complete the game fast. Solvers take longer to complete the game, but do not die as often as veterans. They are adept at solving in-game puzzles. Pacifists are also fast and die often, but they die primarily from active opponents. Runners complete the game quickly, die often, and may ask for help more often from the in-game help-on-demand system.

There are clearly similarities between the models proposed by Bartle [7] and Yee [39]. Figure 1.2 overlays Bartle's [7] interest graph in Fig. 1.1 with Yee's [39] motivation subcomponents. We see that corresponding motivation subcomponents emerge for all four player types. However, Yee [39] also identifies other motivations for play not captured by Bartle's [7] model, in particular motivation for escapism and relaxing.

Yee [39] also argues that empirical evidence suggests player motivations do not suppress each other in the way Bartle [7] assumes. Players may demonstrate subcomponents of more than one of the three overarching motivation components for achievement, social and immersion. Gender and age are two demographic variables that play a role in determining which subcomponents of motivation a player might exhibit.

There are also some broad correlations between the work of Bartle [7] and Yee [39] and the player types identified by Drachen et al. [11]. Solvers have some of the characteristics of Bartle's [7] explorers, while veterans are reminiscent of the achievers identified by Bartle [7] and the mechanics subcomponent of achievement motivation described by Yee [39]. Overall, the player types defined by Drachen et al. have a heavy focus on the 'interest in the world' side of Bartle's interest graph,

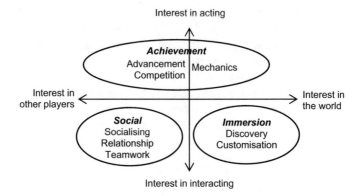

Fig. 1.2 An overlay of a subset of Yee's [39] subcomponents of motivation on Bartle's [7, 8] interest graph. Additional unclassified subcomponents proposed by Yee are 'escapism' and 'relaxing'

as shown in Fig. 1.3. Their metrics have a focus on the proficiency and speed aspects of play, over social and exploratory behaviour.

In another study titled *Coming of Age in Second Life* [9], author Tom Boellstorff conducted an anthropological study of 'life' in the *Second Life* virtual world. By applying the idea of culture as a virtual entity, he uses the methods of a traditional anthropologist to examine issues of gender, race, sex, economics, conflict and antisocial behaviour in the virtual world. Like Bartle [7] and Yee [39], Boellstorff [9] identifies socialising as a key factor in motivating people to use virtual worlds.

Second Life is a nongaming, 'infinite resource' world. However, Boellstorff [9] is also able to identify an analogue to Bartle's [7] 'killer' player type, which he refers to as 'griefers'. Griefers interfere with and degrade the experience of other players, either through socially inappropriate behaviour, or by attacking the

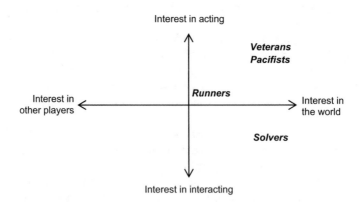

Fig. 1.3 An overlay of a subset of the player types proposed by Drachen et al. [11] on Bartle's [7, 8] interest graph

computational infrastructure of the world itself. People also use *Second Life* to conduct businesses, offer religious teachings or conduct secular education. In these goals, it is perhaps possible to see elements of achievement motivation, although Boellstorff [9] does not classify them as such.

The recurring themes in the gaming and virtual world literature naturally suggest a common underlying model of human behaviour. In the next section we take a closer look at the psychological theories of motivation in the quest for further insights into this underlying model.

1.3 Incentive-Based Theories of Motivation

As we have seen, previous studies have used case-based [7], empirical [39], machine learning [11] and anthropological [9] methodologies for examining and explaining emergent behaviour in online virtual worlds or games. The discussion in this section complements these studies by considering the field of psychological motivation theory [16] to seek explanations for different player types and interactions. The field of motivation is broad, including neuroscientific [19], biological [27], and psychological [16] theories. The review in this section focuses specifically on a subset of motivations that can be considered with reference to incentive. We choose this area because of the natural mapping and relevance of these theories to the player types identified by other researchers. They thus provide a starting point for realising these types as computational models that can be embedded in game-playing agents.

Section 1.3.1 provides a definition of incentive. We then study three motives in particular from the perspective of incentive: achievement motivation in Sect. 1.3.2, affiliation motivation in Sect. 1.3.3 and power motivation in Sect. 1.3.4. These motives have been considered particularly influential by psychologists [16], and form the basis of theories such as the three needs theory [23] and three factor theory [36].

1.3.1 Incentive

Incentives are situational characteristics associated with possible satisfaction of a motive. They are subjective phenomena perceived and affectively evaluated by an individual [16]. A stimulus that triggers a positive affective response in one person may trigger a negative response in another person as a result of their previous experiences. Incentives can be either implicit or explicit. Examples of explicit incentives include money or points in a game. Examples of implicit incentives include challenges to personal control in a performance situation, opportunities for social closeness and opportunities for social control.

Not all motivation theories incorporate a concept of incentive. In addition, motivations that can be considered from a perspective of incentive may also be explained in different ways. This is, in fact, the case with the three motivations described in the next sections. We take an incentive-based view of these theories in this book because we find it particularly conducive to representation in a mathematical (and thus computational) manner.

1.3.2 Achievement Motivation

Achievement motivation drives humans to strive for excellence by improving on personal and societal standards of performance. Achievement motivation is based on estimations of success probability and the difficulty of achieving a goal.

An example of a psychological model of achievement motivation is Atkinson's risk-taking model (RTM) [6]. The RTM was designed to predict individuals' preferences for accepting difficult goals. It defines achievement motivation in terms of conflicting desires to approach success or avoid failure. Six variables are used:

- Incentive for success, I^s, which is equated with value of success,

- Probability of success, P^s, which is equated with difficulty of a task or goal,

- Strength of motivation to approach success, M^s,

- Incentive for avoiding failure, I^f,

- Probability of failure, P^f,

- Strength of motivation to avoid failure, M^f.

The resulting tendency for achievement motivation T_{ach}^{res} is defined as:

$$T_{ach}^{res} = M^s I^s P^s + M^f I^f P^f \tag{1.1}$$

The probabilities of success, P^s, and failure, P^f, have a value between zero (no chance of success/failure) and one (guaranteed success/failure). One of these two outcomes (success or failure) must occur, that is:

$$P^s + P^f = 1 \tag{1.2}$$

Individuals for whom $M^s > M^f$ are termed 'success-motivated', meaning that they tend to formulate approach goals that they achieve by minimizing the difference between their current state and the goal state. Individuals for whom $M^f > M^s$ are termed 'failure-motivated', meaning that they tend to formulate avoidance goals that they achieve by maximizing the difference between their current state and the goal state. These individuals tend to value success less than success-motivated individuals, and consequently tend to underperform their success-motivated colleagues.

Atkinson's model further assumes inverse linear relationships between probability of success and incentive. These relationships reflect everyday experience and empirical data indicating that the feeling of success increases as the perceived probability of success decreases. That is, success at a hard task is worth more than success at an easy task. For success-motivated individuals, the relationship between probability of success and incentive is modelled as:

$$I^s = 1 - P^s \tag{1.3}$$

For failure-motivated individuals, incentive for avoiding failure is highest for very easy tasks. That is, failure at an easy task is worse than failure at a hard task. This can be modelled as:

$$I^f = -P^s \tag{1.4}$$

When the various assumptions are substituted into Eq. 1.1, the tendency for achievement motivation simplifies to:

$$T^{res}_{ach} = (M^s - M^f)(P^s - [P^s]^2) \tag{1.5}$$

Two examples of Eq. 1.5 are shown in Fig. 1.4. Figure 1.4a shows that for success-motivated individuals with $M^s > M^f$, the resultant motivational tendency peaks at a moderate probability of success. This is called the point of maximum approach. The RTM thus implies that success-motivated individuals will select goals of moderate difficulty. In contrast, Fig. 1.4b shows that for failure-motivated individuals with $M^f > M^s$, resultant motivation is minimal for goals with a moderate probability of success. This is called the point of maximum avoidance. Atkinson concludes that failure-motivated individuals tend to select either very easy goals, that have a high probability of success, or very difficult goals for which failure can be easily explained.

More recent work that has examined achievement motivation from an approach-avoidance perspective [14] distinguishes between two types of approach goals: performance-approach goals and mastery-approach goals. Mastery-approach goals are grounded in an intrinsic desire to improve one's competence at a task. Mastery-approach goals generally imply either a self-based or a task-based evaluation of one's competence. Achievement in the context of a mastery-approach goal means 'making progress' or learning. In contrast, performance-approach goals are grounded in a desire to demonstrate or prove competence, especially in the presence of an audience. Performance-approach goals generally imply a norm-based evaluation of one's competence, that is, a demonstration of ability relative to that of others. Achievement in the context of a performance-approach goal means doing better than others. Similarly, two types of avoidance goals are hypothesized: mastery-avoidance and performance-avoidance goals. Achievement in the context of mastery-avoidance means not doing worse than one has done before. Achievement in the context of a performance-avoidance goal means not doing worse than others.

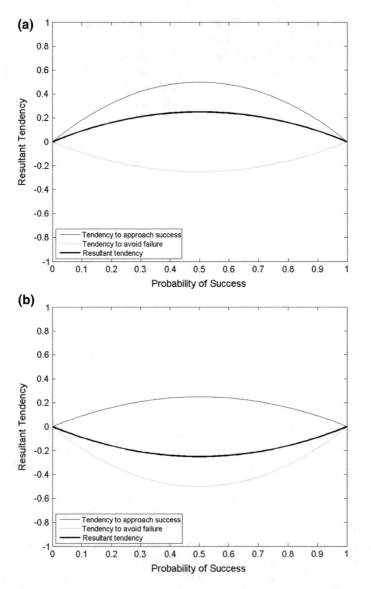

Fig. 1.4 Resultant tendency for motivation according to Eq. 1.5 for **a** an individual motivated to approach success ($M^s = 2$, $M^f = 1$) and **b** and individual motivated to avoid failure ($M^s = 1$, $M^f = 2$). Images from [29]

Approach-avoidance motivation has also been studied in the social domain [32]. In this domain it is used to model the differences in goals concerned with positive social outcomes—such as affiliation and intimacy—and goals concerned with negative social outcomes—such as rejection and conflict [13]. It is understood that the idea of approach-avoidance motivation, along with the concepts of incentive and probability of success, are particularly important not only for achievement motivation, but also for affiliation, power and other forms of motivation [16]. The following sections thus discuss affiliation and power motivation with reference to approach-avoidance theory, incentive and probability of success. These concepts are then used as the basis for the computational models for achievement, affiliation and power presented in Chap. 2.

1.3.3 Affiliation Motivation

Affiliation refers to a class of social interactions used by individuals to seek contact with other, formerly unknown or little known, individuals and maintain contact with those individuals in a manner that both parties experience as satisfying, stimulating and enriching [16]. The need for affiliation is activated when an individual comes into contact with another unknown or little known individual.

As with achievement motivation, affiliation motivation is thought to comprise of two contrasting components: hope of affiliation and fear of rejection. When unfamiliar people interact, the hope component is activated first. Under the influence of affiliation motivation, contact is initiated. As familiarity with the person increases, the closer the relationship becomes and the more painful it would be if rejection occurred. The fear of rejection is activated and becomes increasingly strong. Sensitivity to relevant signals is heightened until the point of maximum conflict between approach and avoidance. When fear becomes dominant the closeness of the relationship is diminished until the fear decreases, hope dominates once again and the cycle begins anew.

The maximum approach-avoidance conflict occurs at the point where both components are equally strongly aroused. Although the avoidance tendency is activated later, the gradient of avoidance is steeper than the gradient of approach. Specific affiliation-related goals might include being in the company of others, cooperating, exchanging information and being friends.

Individuals high in affiliation motivation may also be intent on effecting reconciliation with others, may make more suggestions to change the attitudes of others to bring those attitudes more into line with their own, may avoid games of chance and may initiate fewer acts that might spark conflict. This can mean that they also initiate less cooperative acts. Individuals with medium to high affiliation motivation may be less deceptive than those with low affiliation motivation.

In addition to these specific affiliation-related goals, strength of affiliation motivation can also influence the way that individuals assess other goals. While theories of affiliation have not been developed mathematically to the extent of the RTM, affiliation can be considered from the perspective of incentive and probability of success [16]. In contrast to individuals high in achievement motivation, individuals high in affiliation motivation may select goals with a higher probability of success (or lower incentive). This preference for low-risk activities can be understood as a preference to avoid public competition and conflict by avoiding goals that can lead to the acquisition of resources or reinforcers that are desirable to others. This trend was identified in experiments with participants playing rounds of roulette in casino settings [24]. When participants placed their bets and won or lost in front of the whole group, affiliation-motivated individuals showed a significant tendency to make low-risk bets. This is in contrast to power-motivated individuals, who made more high-risk bets. Heckhausen and Heckhausen [16] thus identify affiliation motivation as an important counterbalance to power motivation.

1.3.4 Power Motivation

Power can be described as a domain-specific relationship between two individuals, characterized by the asymmetric distribution of social competence, access to resources or social status [16]. Power is manifested by unilateral behavioural control and can occur in a number of different ways. Suppose there are two individuals, A^1 and A^2:

- *Reward power* is exerted if A^1 is in a position to satisfy a motive of A^2 and makes such satisfaction contingent on the behaviour of A^2;

- *Coercive power* is exerted if A^1 is in a position to punish a behaviour by A^2 by withdrawing their opportunity to satisfy certain motives. Furthermore, A^1 makes such punishment contingent on the behaviour of A^2;

- *Legitimate power* is derived from norms internalized by A^2 that tell A^2 that A^1 is authorized to regulate their behaviour;

- *Referent power* arises from a desire of A^2 to be like A^1;

- *Expert power* is determined by the extent to which A^2 perceives A^1 as having special knowledge or skills in a particular area;

- Informational power is exerted when A^1 communicates information to A^2 that triggers A^2 to change their beliefs and behaviour.

Five components of fear (avoidance) of power have also been identified: fear of the augmentation of one's power source, fear of the loss of one's power source, fear of exerting power, fear of the counter-power of others and fear of one's power behaviour failing. These inhibition tendencies moderate power by channelling the expression of power into socially acceptable behaviour.

Like affiliation, power motivation can be considered with respect to incentive and probability of success. There is evidence to indicate that the strength of satisfaction of the power motive depends solely on incentive and is unaffected by the probability of success of a goal [24]. This trend was identified in the same experiments discussed previously, with participants playing rounds of roulette in casino settings. When participants placed their bets and won or lost in front of the whole group, power-motivated participants showed a preference for the highest-risk bets. Power-motivated individuals select high-incentive goals, as achieving these goals gives them significant control of the resources and reinforcers of others.

Power motivation has a number of potential roles in future self-motivated agents. In particular, it plays an important role in setting a system, whether an individual or a society, into 'expansion mode'. This is because risk-taking behaviour is necessary for both establishing boundaries and identifying and exploiting high return situations. In economic settings, for example, individuals with high power motivation coupled with high achievement motivation and low affiliation motivation are frequently found in management roles of successful or expanding businesses. This combination of motives has been labelled the leadership motive profile (LMP) [25]. At a national level, high power motivation combined with low achievement and affiliation motivation heightens the risk of nations entering into armed conflict to expand their international influence, resources or territory. This combination of motives has been labelled the imperial motive profile [25].

1.3.5 Dominant Motives and Motive Profiles

McClelland [23] writes that, regardless of gender, culture or age, an individual's implicit motive profile tends to have a dominating motivational driver. That is, one of the three motives discussed above will have a stronger influence on decision making than the other two, but the individual will not be conscious of this. The dominant motive might be a result of cultural or life experiences and results in distinct individual characteristics. Some of these are summarised in Table 1.2.

Hybrid profiles of power, affiliation and achievement motivation have also been associated with distinct individual characteristics. For example, there appears to be a relationship between certain combinations of dominant and non-dominant

Table 1.2 Characteristics that may be observed in individuals with a given dominant motivational driver [16, 23, 30]

Dominant motivational driver	Characteristics
Achievement	• Prefers choosing and accomplishing challenging goals • Willing to take calculated risks • Likes regular feedback • Often likes to work alone
Affiliation	• Wants to belong to a group • Wants to be liked • Prefers collaboration over competition • Does not like high risk or uncertainty
Power	• Wants to control and influence others • Likes to win • Likes competition • Likes status and recognition

motives and the emergence of leadership abilities in an individual. In one study of leadership [26], the careers of 237 entry level managers were studied over a 16 year period. The study hypothesized that managers with higher power motivation than affiliation motivation would be more successful in their careers. It was shown that managers whose motives follow the LMP were more likely to be promoted to a higher level in their careers. On the other hand, those with high achievement motivation were more likely to be successful at lower management levels where technical skills are needed, and less successful at higher management levels. These findings have been used to develop LMP theory [22].

LMP theory describes and predicts leadership effectiveness and success according to an individual's strength of affiliation, achievement and power motivation [26]. McClelland and Watson [24] further emphasize that to become an effective leader, a moderate strength of achievement is required. Achievement motivation is required for leadership, but stronger achievement than power motivation may encourage personalized power. People who are high in achievement motivation tend to pursue individual achievement. In this case, there is a danger of the emergence of personalized power where personal gain is prioritized at the expense of others [26].

LMP theory [22] characterises a leader as having weaker affiliation motivation. This is because individuals who have a strong affiliation motivation tend to prioritize relationships with others and are more likely to avoid conflicts. However, leadership, which involves influencing others, may out of necessity require conflict.

The remainder of this book focuses primarily on modelling profiles with a single dominant motive, but there is of course potential for future work to look beyond this.

1.3.6 Motivation and Zeitgeist

As intimated above, motivation theorists also understand that the proportions of individuals with different motives in a population contribute to a 'national Zeitgeist' [25]. Literally, this translates as a national 'spirit of the age' or national 'spirit of the time'. The national Zeitgeist is the dominant set of preferences that influence the culture during a particular period.

McClelland linked Zeitgeist and motivation through the concept of a national motive index (NMI). The NMI measures the extent to which the relative strength of the power and affiliation imagery prevailing in a country are a portent of war or national unrest. The focus is on power and affiliation motivation in this case (and not on achievement motivation), because strong power motivation is associated with competition and conflict, while affiliation motivation is understood to be a counterbalance to power motivation. The NMI is calculated by analysing the content of texts that are popular during the time period of interest. This includes children's books, popular novels and songs.

In the same way McClelland argued for a national Zeitgeist, we saw earlier in this chapter that Boellstorff argues for the existence of virtual world culture. In fact, many modern MMORPGs offer different instances of the game world that support different levels of competitive game play. Competitive player-versus-player (PVP) game play is supported by PVP servers. Player-versus-environment (PVE) servers permit players to focus on cooperative social interactions with other players. We thus hypothesise that the concept of a NMI has a parallel in virtual world and computer game settings. We consider this idea further in Chap. 8.

1.4   Conclusion

In Sect. 1.2 we identified similarities between Bartle's [7, 8] player types, Yee's [39] motivation components and different cultural and community groups identified by Boellstorff [9]. We now propose a mapping between these groups and the 'three needs' [23] or 'influential trio' [16] of achievement, affiliation and power motives. We discuss these once again with reference to Bartle's interest graph in Fig. 1.5.

First, there are similarities between the killer player type described by Bartle, the advancement and competition subcomponents of motivation identified by Yee [39], the griefers identified by Boellstorff [9] and the resource-controlling desires of a power-motivated individual. Figure 1.5 thus places power-motivated individuals in the upper left quadrant, corresponding to interest in acting on other players.

Likewise, the concept of socialiser, socialising or affiliation is prevalent among the conclusions of different researchers. Figure 1.5 places affiliation-motivated individuals in the bottom left quadrant, corresponding to interest in interacting with other players.

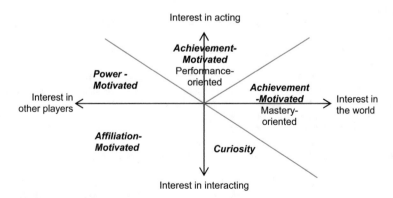

Fig. 1.5 Psychological motivation on Bartle's [7, 8] interest graph. The *right-hand side* of the interest graph is dominated by achievement motivation. The *left-hand side* includes power- and affiliation-motivated individuals

A clear concept of achievers or achievement emerges from the literature in both the gaming and motivation psychology domains. A performance-oriented view of achievement motivation sits towards the centre of Bartle's player-world spectrum (horizontal axis) because players evaluate their achievements against other players. Hybrid profiles such as the LMP also sit in this area. In contrast, a mastery-oriented view of achievement sits further to the right because players evaluate their competence in the context of a task only. Thus, achievement motivation conceivably plays a role across three quadrants of Bartle's interest graph.

Finally, there are similarities between the subcomponents of the immersion category proposed by Yee [39] and the explorer player type hypothesised by Bartle [7, 8]. Some of these behaviours can be explained in terms of mastery-oriented achievement motivation. Others are well explained by other kinds of intrinsic motives such as curiosity and novelty-seeking behaviour. Both of these are beyond the scope of this book (see [28] for a review).

We conclude that theories of motivation such as achievement, affiliation and power are complementary to our understanding of player types and provide us with a well-researched starting point for developing computational models of motivation that have the potential to exhibit some of the diverse game-playing characteristics of humans in virtual worlds. The development of such models is the topic of Chap. 2.

References

1. Ultima Online, http://www.uo.com/ *Broadsword Online Games,* 1997
2. EverQuest, https://www.everquest.com *Sony Online Entertainment,* 1999
3. Star Wars Galaxies, *Sony Online Entertainment,* 2003
4. Tomb Raider: Underworld, http://www.tombraider.com/ *Crystal Dynamics,* 2008
5. S. Arnab, K. Brown, S. Clarke, I. Dunwell, T. Lim, N. Suttie, S. Louchart, M. Hendrix, S. de Freitas, The development approach of a pedagogically-driven serious game to support relationship and sex education (RSE) within a classroom setting. Comput. Educ. **69**, 15–30 (2013)
6. J.W. Atkinson, Motivational determinants of risk-taking behavior. Psychol. Rev. **64**, 359–372 (1957)
7. R. Bartle, Hearts, clubs, diamonds, spades: Players who suit MUDs. J. Virtual Environ **1**, (online) (1996)
8. R. Bartle, *Designing Virtual Worlds* (New Riders, Indianapolis, 2004)
9. T. Boellstorff, *Coming of Age in Second Life: An Anthropologist Explores the Virtually Human* (Princeton University Press, Princeton, New Jersey, 2008)
10. I. Briggs Myers, P. Myers, *Gifts Differing: Understanding Personality Type* (Davies-Black Publishing, Mountain View, CA, 1995)
11. A. Drachen, A. Canossa, G. Yannakakis, Player modeling using self-organisation in Tomb Raider: Underworld, in *Proceedings of the IEEE Symposium on Computational Intelligence and Games*, 2009, pp. 1–8
12. A. Drachen, C. Thurau, J. Togelius, G. Yannakakis, C. Bauckhage, Game data mining, in *Game Analytics: Maximising the Value of Player Data*, ed. by M. Seif El-Nasr, A. Drachen, A. Canossa (Springer-Verlag, London, 2013)

13. A. Elliot, *Handbook of Approach and Avoidance Motivation* (Taylor and Francis, New York, NY, 2008)
14. A. Elliot, A. Eder, E. Harmon-Jones, Approach-avoidance motivation and emotion: Convergence and divergence. Emot. Rev. **5**, 308–311 (2013)
15. Flatfingers, *Competition and cooperation in MMORPGs* http://flatfingers-theory.blogspot. com.au/2008/07/competition-and-cooperation-in-mmorpgs.html. Accessed 8 July 2008
16. J. Heckhausen, H. Heckhausen, *Motivation and Action* (Cambridge University Press, New York, NY, 2010)
17. M. Hendrix, S. Meijer, J. Van Der Velden, A. Iosup, Procedural content generation for games: A survey. ACM Trans. Multimedia Comput. Commun. Appl. **9**, 1–22 (2013)
18. J. Juul, *A Casual Revolution: Reinventing Video Games and Their Players* (The MIT Press, Cambridge, MA, 2009)
19. R. K and W. Lee, Neuroscience and human motivation, in *The Oxford Handbook of Human Motivation*, ed. by R. Ryan (Oxford University Press, Inc, New York, 2012)
20. J. Laird, M. van Lent, Interactive computer games: Human-level AI's killer application, in *Proceedings of the National Conference on Artificial Intelligence (AAAI)*, 2000, pp. 1171–1178
21. P. Lameras, P. Petridis, K. Torrens, I. Dunwell, M. Hendrix, S. Arnab, Training science teachers to design inquiry-based lesson plans through a serious game, in *Proceedings of the Sixth International Conference on Mobile, Hybrid and Online Learning*, 2014, pp. 86–91
22. R. Lussier, C. Achua, *Leadership: Theory, Application and Skill Development* (South Western, Cengage Learning, Mason, OH, 2012)
23. D. McClelland, *The Achieving Society* (The Free Press, New York, NY, 2010)
24. J. McClelland, R.I. Watson, Power motivation and risk-taking behaviour. J. Pers. **41**, 121–139 (1973)
25. J. McClelland, *Power: The Inner Experience* (Irvington, New York, 1975)
26. J. McClelland, R. E. Boyatzis, The leadership motive pattern and long term success in management. J. Appl. Psychol. **39** (1982)
27. D. McFarland, *Animal Behaviour: Psychobiology, Ethology and Evolution* (Pearson Education Limited, Harlow, England, 1999)
28. K. Merrick, M.L. Maher, *Motivated Reinforcement Learning: Curious Characters for Multiuser Games* (Springer, Berlin, 2009)
29. K. Merrick, K. Shafi, Achievement, affiliation and power: Motive profiles for artificial agents. Adapt. Behav. **19**, 40–62 (2011)
30. K. Merrick, The role of implicit motives in strategic decision-making: Computational models of motivated learning and the evolution of motivated agents. GAMES, Special Issue on Psychological Aspects of Strategic Choice **6**, 604–636 (2015)
31. D. Merrill, R. Reid, *Personal Styles and Effective Performance* (CRC Press, New York, 1999)
32. J. Nikitin, A. Freund, When wanting and fearing go together: The effect of co-occurring social approach and avoidance motivation on behavior, affect and cognition. Eur. J. Soc. Psychol. **40**, 783–804 (2009)
33. C. Pedersen, J. Togelius, G. Yannakakis, Modeling player experience in Super Mario Bros., in *Proceedings of the IEEE Symposium on Computational Intelligence and Games*, Milan, Italy, 2009, pp. 132–139
34. J. Reahard, Second Life readies for 10th anniversary, celebrates a million active users per month (2013) *Massively by Joystiq* http://massively.joystiq.com/2013/06/20/second-life-readies-for-10th-anniversary-celebrates-a-million-a/. Accessed 30 Aug 2014
35. R. Saunders, J.S. Gero, The digital clockwork muse: A computational model of aesthetic evolution," in *Proceedings of the AISB'01 Symposium on AI and Creativity in Arts and Science, SSAISB* (2001)
36. D. Sirota, L. Mischkind, M. Meltzer, *The Enthusiastic Employee* (Pearson Education Inc, Upper Saddle River, NJ, 2005)

37. S. Tognetti, M. Garbarino, A. Bonarini, M. Matteucci, Modeling enjoyment preference from physiological responses in a car racing game, in *Proceedings of the IEEE Conference on Computational Intelligence and Games*, Copenhagen, Denmark, 2010, pp. 18–21
38. G. Yannakakis, Game AI revisited, in *Proceedings of the Ninth Conference on Computing Frontiers*, Cagliari, Italy, 2012, pp. 285–292
39. N. Yee, Motivations of play in online games. Cyberpsychol. Behav. **9**, 772–775 (2007)

Chapter 2
Computational Models of Achievement, Affiliation and Power Motivation

Chapter 1 examined literature from motivation psychology and reviewed the theories that may contribute to different game-play characteristics in humans. It specifically focused on three theories of motivation that can be modelled using the concept of incentive: achievement, affiliation and power motivation. This chapter introduces computational models of motivation to embed these human-inspired motives in artificial agents. A flexible mathematical model is introduced that permits these three motives to be expressed in terms of approach and avoidance components, which can be adjusted to create different motivation variants. Two approaches to using these models for goal selection are introduced.

2.1 Towards Computational Motivation

In the previous chapter we identified a mapping between three psychological motivation theories for achievement, affiliation and power and the player types, motivation components and user groups that have been identified among participants in virtual worlds. We described how theories of implicit motivation proposed by motivation psychologists can explain differences in behaviour between individuals seemingly in the same situation. This suggests a methodology for developing more diverse and believable computer-controlled game characters by embedding them with computational models of those motivations.

This chapter presents computational models of motivation for achievement, affiliation and power motivation. The models are designed such that they can be used in isolation or together, embedded in an artificial 'motive profile'. This chapter draws on the ideas of incentive, probability of success and approach-avoidance motivation seen in Chap. 1 as the basis for developing computational models for achievement, affiliation and power motivation.

Our basic approach uses sigmoid curves to model approach and avoidance of a goal as a function of either the probability of successfully achieving the goal, or the

© Springer International Publishing AG 2016
K.E. Merrick, *Computational Models of Motivation for Game-Playing Agents*,
DOI 10.1007/978-3-319-33459-2_2

incentive to achieve a goal. A goal here is an intermediate construct that specifies how an individual strategically goes about addressing the underlying approach and avoidance motives [7]. In artificial agent research, goals can describe states or changes to achieve, states to maintain or preserve, information to retrieve and behaviours to execute or cease, among other things [3]. We do not, at this point fix on a specific definition for a goal, but note that it could be any of these things in an artificial agent. We revisit the idea of a goal later in the book when we embed computational models of motivation in specific agent architectures.

Section 2.2 introduces a set of mathematical tools we can use to model approach-avoidance motivation. These are then utilised to produce three incentive-based computational models of achievement, affiliation and power motivation. These motives are considered individually and in combination in a computational motive profile in Sect. 2.3. Models of different complexity and fidelity are considered. Finally, Sect. 2.4 presents two approaches to goal selection using motivation.

2.2 Modelling Incentive-Based Motives Using Approach-Avoidance Theory

The approach-avoidance theory of motivation is characterised by the idea that both the attractiveness and repulsiveness of a goal or an incentive increase the closer one gets to it. Closeness here may refer to closeness in time, space or psychological distance. Different representations of this theory have been proposed. Miller [19], for example, focused on the gradient of approach and avoidance. He hypothesised that the strength of avoidance motivation may increase more rapidly than the strength of approach motivation as the goal or incentive is neared. A typical linear model of this phenomenon has the general form:

$$y = mx + b, \tag{2.1}$$

where x is the distance to incentive, y is the strength of the resultant tendency of motivation for the incentive, m is a negative number controlling the gradient of approach or avoidance and b is a positive number controlling the maximum strength of the motivation in question. An example is shown in Fig. 2.1.

An alternative model proposed by Maher [12] focuses on the strength of motivation in approach-avoidance conflict, rather than on the gradients of approach and avoidance. When an individual is far from a goal or incentive, the motivation to avoid it is greater than the motivation to approach it. However, when the individual approaches the incentive, the strength of motivation to approach the incentive can become very strong very quickly. The model has hyperbolic characteristics of the following form:

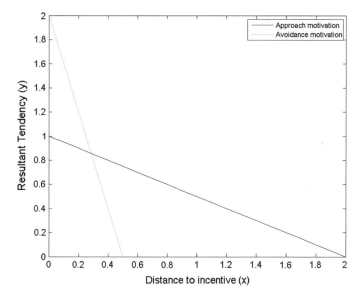

Fig. 2.1 Modelling approach-avoidance theory with a focus on gradient [19] using Eq. 2.1 with $m = -0.5$, $b = 1$ for approach motivation and $m = -4$, $b = 2$ for avoidance motivation

$$y = \frac{a}{bx} + c. \tag{2.2}$$

x is again the distance to incentive and y is the strength of the resultant tendency of motivation for the incentive. The parameters a and b control the gradient of approach or avoidance and c controls the minimum strength of the motivation in question. An example is shown in Fig. 2.2.

Mathematically, we can see that the two models are different, but do retain some similarities. In particular, the strength of motivational tendency for both approach and avoidance increases as distance to incentive decreases.

The models above consider motivational tendency as a function of distance to incentive. Another key concept of motivation is the U- or inverted-U-shaped relationship between motivational tendency and strength of incentive. This relationship is captured, for example, by the quadratic model of the resultant tendency curves for motivation in Eq. 1.5. A more general form for this model is (Fig. 2.3):

$$y = av^2 + bv + c \tag{2.3}$$

The variable v is the value of the incentive. The parameters a and b together control the gradient of approach and avoidance of incentive, the position of the maximum (or minimum) motivational tendency along the horizontal axis, and the strength of the maximum (or minimum) motivational tendency. c controls the intersection point of the resultant motivational tendency curve with the vertical axis.

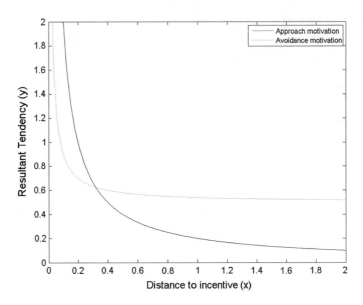

Fig. 2.2 Modelling approach-avoidance theory with a focus on strength [12] using Eq. 2.2 with $a = 0.2$, $b = 1$, $c = 0$ for approach motivation and $a = 0.2$, $b = 5$, $c = 0.5$ for avoidance motivation

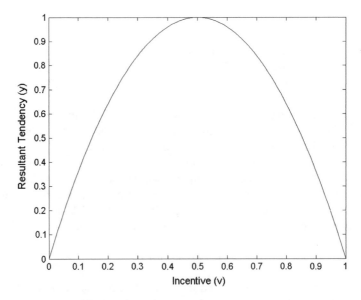

Fig. 2.3 A quadratic model of motivation as a function of incentive using Eq. 2.3 with $a = -4$, $b = 4$, $c = 0$. We assume incentive is inversely proportional to probability of success

A number of other mathematical functions have been used to model phenomena with similar inverted-U-shaped characteristics, such as arousal, hedonic response and creativity [16]. One example is a Gaussian function of the form:

$$y = ae^{-\frac{(v-b)^2}{2c}} \tag{2.4}$$

This function has parameters a, b and c that give us control of the height of the inverted U, the position of the maximum and the gradient of increase and decrease of the function. The variable v here can represent factors such as the amount of neural activity, psychophysical intensity, ecological stimuli or collative effects such as novelty or incentive, depending on the phenomenon being modelled [9]. Using this model, the parameter b effectively controls the gradient of both approach and avoidance, so we cannot manipulate these independently (Fig. 2.4).

When modelling hedonic response as the sum of positive and negative feedback for approach and avoidance, a Gaussian cumulative distribution can be used to model the positive feedback as the area under the Gaussian probability distribution. However, other functions are also common as cumulative distribution functions, and can give us greater control of the shape of the motivation curve. One example is a sigmoid function:

$$y = \frac{a}{1 + e^{-c(v-b)}} \tag{2.5}$$

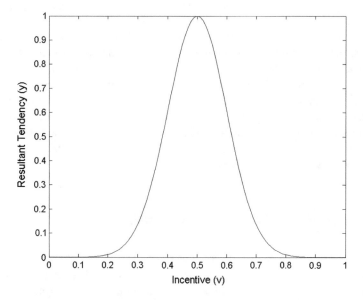

Fig. 2.4 Gaussian model of motivation as a function of incentive using Eq. 2.4 with $a = 1$, $b = 0.5$, $c = 0.01$

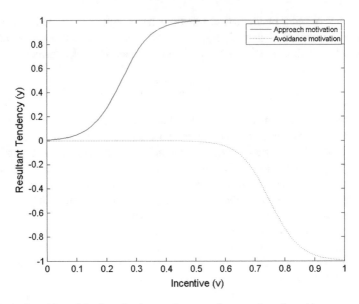

Fig. 2.5 A sigmoid model of motivation as the sum of approach and avoidance curves using Eq. 2.5 with $a = 1$, $b = 0.25$, $c = 20$ for approach motivation and $a = -1$, $b = 0.75$, $c = 20$ for avoidance motivation

a again controls the maximum of the function and c the rate of increase. However, b now controls the position of the turning point of the curve along the horizontal axis. v is again the amount of neural activity, psychophysical intensity, ecological stimuli or collative effect being modelled (Fig. 2.5).

The inverted-U-shaped relationship between probability of success and resultant tendency for motivation can be achieved using a sigmoid-based representation with one sigmoid function for positive feedback (approach) and another for negative feedback (avoidance) [18]. This effectively gives us separate parameters to control the strength, rate of increase and position of the turning points for approach and avoidance motivation. A sigmoid representation has previously been used in this way to model curiosity and interest [22] as approach to novelty and avoidance of very high novelty. We use these ideas to model achievement, affiliation and power motivation in the following sections.

2.2.1 Modelling Achievement Motivation

Atkinson's risk-taking model (RTM; see Chap. 1) has been both influential and successful in aiding the understanding of achievement motivation in humans. The general trends described by the RTM have been observed in experimental settings in humans [2, 13]. However, the point of maximum approach tends to fall below the critical level of $P^s = 0.5$ predicted by the RTM. Furthermore, failure-motivated individuals do not select extremely difficult goals to the extent predicted by the RTM. The known limitations of the RTM suggest that a more sensitive model is required to capture the subtleties of achievement motivation in artificial agents.

The ideas of incentive, probability of success and approach-avoidance motivation proposed by Atkinson can be captured in a sigmoid-based model. Thus, such a model does not redefine the existing psychological model of motivation, but rather interprets it computationally in a flexible manner that can potentially be extended to other approach-avoidance motivations.

Equation 2.6 represents achievement motivation as the difference between two sigmoid functions for approach and avoidance of a goal G. Using a sigmoid representation, approach motivation is stronger for goals with a higher probability of success, until a certain threshold probability is reached and approach motivation plateaus.

Conversely, avoidance motivation is zero for goals with a very low probability of success, and negative for goals with a high probability of success. This means that failure at a very easy goal is punished the most. The resultant tendency to achieve a goal G is the sum of the approach and avoidance sigmoid curves in composition with a function $P^s(G)$ for computing the subjective probability of successfully achieving goal G, as shown in Eq. 2.6:

$$T_{\text{ach}}^{\text{res}}(P^s(G)) = \left(T_{\text{ach}}^{\text{res}} \circ P^s\right)(G) = \frac{S_{\text{ach}}}{1 + e^{-\rho_{\text{ach}}^+(P^s(G)-M_{\text{ach}}^+)}} - \frac{S_{\text{ach}}}{1 + e^{-\rho_{\text{ach}}^-(P^s(G)-M_{\text{ach}}^-)}}.$$

$$(2.6)$$

Equation 2.6 is visualized in Fig. 2.6. $P^s(G)$ has a value range between zero (guaranteed failure) and one (guaranteed success). The manner in which probability of success is estimated will influence the resulting achievement motivation value computed. Various methods have been proposed by psychologists. These include the mastery- and performance-oriented approaches summarised in Table 2.1. Each of these approaches is also possible for artificial agents that can interact with their environment or other agents. Self-based estimates and social comparison standards are perhaps the most straightforward, as they can be understood as the number of successful attempts divided by the total number of attempts.

Measuring probability of success in absolute standards is more difficult as it requires an understanding of the domain in question.

The model has five parameters: M_{ach}^+, M_{ach}^-, ρ_{ach}^+, ρ_{ach}^- and S_{ach}, which are summarized in Table 2.2. M_{ach}^+ is the turning point of the sigmoid for approach motivation and M_{ach}^- is the turning point of the sigmoid for avoidance. When the approach turning point is to the left of the avoidance turning point (that is, $M_{\text{ach}}^+ < M_{\text{ach}}^-$), then the resultant tendency represents a success-motivated individual, as shown in Fig. 2.6a. Note the characteristic inverted U-shape of the curve for resultant tendency, similar to that seen in Atkinson's [1] model in Fig. 1.4a.

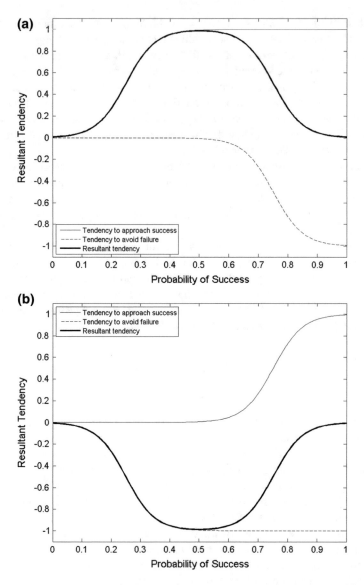

Fig. 2.6 Sigmoid representations of **a** motivation to approach success ($S_{ach} = 1$, $M_{ach}^+ = 0.25$, $M_{ach}^- = 0.75$ and $\rho_{ach}^+ = \rho_{ach}^- = 20$) and **b** motivation to avoid failure ($S_{ach} = 1$, $M_{ach}^+ = 0.75$, $M_{ach}^- = 0.25$ and $\rho_{ach}^+ = \rho_{ach}^- = 20$). Images from [18]

Success-motivated individuals with a lower value of M_{ach}^+, are more likely to attempt goals with a lower probability of success. For these individuals, the lower the value of M_{ach}^-, the smaller the range of success probabilities they will consider to be highly motivating. This means that their behaviour will be more focused on goals within a narrow range of success probabilities.

Table 2.1 Ways in which individuals may estimate probability of success at a goal

Mastery-oriented estimates	Performance-oriented estimates
Self-based estimates from individual experience: how well an individual has performed on previous attempts at the goal	Norm-based estimates or social comparison standards: how many other people can solve the goal (see Chap. 9)
Task-based estimates or absolute standards: for example, distance from a target in a shooting or throwing game (see Chap. 4)	

Table 2.2 Parameters of the achievement motivation model in Eq. 2.6 and their possible values

Parameter	Description	Value range
$P^s(G)$	Probability of success	$[0, 1]$
M_{ach}^+	Turning point of success approach	$(-\infty, \infty)$
M_{ach}^-	Turning point of failure avoidance	$(-\infty, \infty)$
ρ_{ach}^+	Gradient of success approach	$[0, \infty)$
ρ_{ach}^-	Gradient of failure avoidance	$[0, \infty)$
S_{ach}	Motivation strength	$[0, \infty)$

$M_{ach}^+ > M_{ach}^-$ can be used to model Atkinson's [1] original concept of a failure-motivated individual, as in Fig. 2.6b. Note the characteristic U shape of the curve for resultant tendency, similar to that seen shown in Atkinson's [1] model in Fig. 1.4b. In a failure-motivated individual the magnitude of negative feedback (punishment) for failing increases more quickly than in success-motivated individuals. This has a tendency to focus behaviour on very difficult goals with a low probability of success. The positive feedback for success increases slowly, producing a tendency to focus on very easy goals with a high probability of success.

Atkinson and Litwin [2] later identified subtypes of achievement motivation based on the combination of tendency to approach success and tendency to avoid failure. Individuals' tendency to approach success or avoid failure was gauged using the projective test of need achievement and Mandler-Sarason tests. Individuals were then broken into four groups as follows:

- H-L: high motivation to approach success and low motivation to avoid failure,

- H-H: high motivation both to approach success and to avoid failure,

- L-L: low motivation both to approach success and to avoid failure

- L-H: low motivation to approach success and high motivation to avoid failure.

In this study, Atkinson and Litwin [2] found that failure-motivated individuals do not select very easy or very difficult tasks to the extent predicted by their original

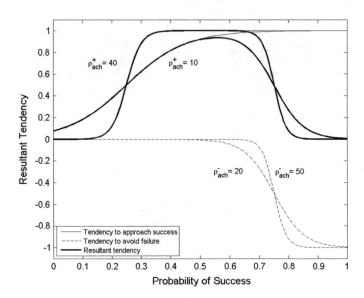

Fig. 2.7 Two examples of motivation to approach success with varying ρ_{ach}^{+} and ρ_{ach}^{-}. $S_{ach} = 1$, $M_{ach}^{+} = 0.25$, $M_{ach}^{-} = 0.75$ and $\rho_{ach}^{+} < \rho_{ach}^{-}$ in both cases

RTM. Rather, they simply have slightly lower tendency for tasks of intermediate difficulty. We model and study these subtypes with this fact in mind in Chap. 4.

Using a sigmoid-based model gives us additional control of the shape of the resultant tendency curve through two parameters controlling the gradients of approach and avoidance. Specifically, $\rho_{ach}^{+} > 0$ is the gradient of approach to success and $\rho_{ach}^{-} > 0$ is the gradient of avoidance of failure. Some experimental evidence with biological motives for hunger satiation and pain avoidance [5] suggests the gradient of approach is often less than the gradient of avoidance. More recent work [8] connects approach-avoidance motivation more broadly to concepts of appetition, reward and incentive (approach) as well as aversion, punishment and threat (avoidance). In the case of achievement motivation, gradient of approach less than gradient of avoidance would imply $\rho_{ach}^{+} < \rho_{ach}^{-}$. Two examples of such scenarios are shown in Fig. 2.7. This figure demonstrates how we can tune the shape of the motivation curve and the impact of changes to ρ_{ach}^{+} and ρ_{ach}^{-} on the shape of the motivation curve.

Finally, S_{ach} determines the strength of achievement motivation. When multiple approach-avoidance motives are modelled using a sigmoid-based approach, the S parameter permits them to compete or cooperate to control the behaviour of the individual. As with the success and failure gradients, the range for S_{ach} is somewhat arbitrary, and depends on the ranges of the S values for other motivations, if any. A general range of parameter values possible in this model is summarized in Table 2.2. As the discussion above suggests, in practice, specific constraints on these ranges create motive profiles that are more or less realistic in human motivational terms. This will be demonstrated in Chap. 4.

2.2.2   Modelling Affiliation Motivation

As we saw in Chap. 1, approach-avoidance motivation theory has been extended to
the social domain [7, 20]. Thus, in this section we model affiliation motivation as
the difference between two sigmoid functions for hope of affiliation and fear of
rejection. Our model also interprets affiliation motivation as a counterbalance to
power motivation. As discussed in Chap. 1, McClelland and Watson [14] presented
evidence indicating that the strength of satisfaction of the power motive depends
solely on incentive and is unaffected by the probability of success. Power-motivated
individuals select high-incentive goals, as achieving these goals gives them significant control of the resources and reinforcers of others. To represent affiliation
motivation as opposing power motivation, we thus define hope and fear of affiliation with respect to incentive. In contrast to power motivation, hope of affiliation
(approach motivation) is high for low-incentive goals. These goals are likely to
cause the least conflict with others by competing for control of their resources or
reinforcers. As goal incentive increases, approach motivation decreases and plateaus, as shown in Fig. 2.8. Negative feedback (avoidance motivation) is greatest
for high-incentive goals which may cause conflict with others. These two sigmoid
curves are summed and composed with goal incentive $I^s(G)$ as shown in Eq. 2.7 to
get the resultant tendency for affiliation, which peaks for low-incentive goals:

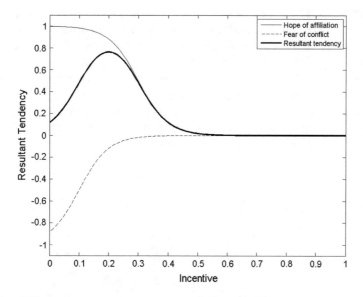

Fig. 2.8 Affiliation motivation as the sum of curves for hope of affiliation and fear of conflict as in
Eq. 2.7 ($S_{\text{aff}} = 1$, $M_{\text{aff}}^+ = 0.3$, $M_{\text{aff}}^- = 0.1$ and $\rho_{\text{aff}}^+ = \rho_{\text{aff}}^- = 20$). Image from [18]

$$T_{\text{aff}}^{\text{res}}(I^s(G)) = \left(T_{\text{aff}}^{\text{res}} \circ I^s\right)(G) = \frac{S_{\text{aff}}}{1 + e^{-\rho_{\text{aff}}^+\left(M_{\text{aff}}^+ - I^s(G)\right)}} - \frac{S_{\text{aff}}}{1 + e^{-\rho_{\text{aff}}^-\left(M_{\text{aff}}^- - I^s(G)\right)}}. \quad (2.7)$$

$I^s(G)$ represents the incentive to complete a given goal G. The process of estimating incentive is imperfectly understood for humans, both in terms of the units in which incentive might be measured and the way goals are mapped to incentive values. In addition there is a great deal of conflicting experimental evidence in this area [10]. In this chapter, incentive is represented as a value between zero and one, with incentive of one denoting the most valuable goals and incentive of zero denoting the least valuable goals. Some possible suggestions for modelling incentive include:

- Incentive inversely proportional to probability of success: this is commonly assumed for achievement goals. It should be noted that the gradient of this mapping can vary from person to person;

- Incentive proportional to explicit value: certain goals directly satisfy motivation, for example, consuming food satisfies hunger;

- Incentive proportional to socially determined value: certain goals have indirect value depending on circumstances. For example, earning money is valuable in a capitalist society.

The first and third approaches are the easiest to model in generic terms. For the first approach, the agent can determine probability of success using one or more of the approaches discussed in the previous section. For the third approach the agents can communicate to agree on goal values. The second approach is the most difficult as it implies some domain knowledge of which objects are inherently valuable. This could be achieved through learning or exploration strategies such as trial-and-error.

This model also has five parameters, M_{aff}^+, M_{aff}^-, ρ_{aff}^+, ρ_{aff}^- and S_{aff}, which are summarized in Table 2.3. M_{aff}^+ is the turning point of the curve describing the approach component of affiliation motivation and M_{aff}^- is the turning point of the curve describing the avoidance component. For affiliation motivation there is the constraint that hope of affiliation should drop in response to an increase in fear of rejection. This means that $M_{\text{aff}}^+ > M_{\text{aff}}^-$ is required.

Table 2.3 Parameters of affiliation motivation (Eq. 2.7) and their possible values

Parameter	Description	Value range
$I^s(G)$	Incentive value for success at goal G	$[0, 1]$
M_{aff}^+	Turning point of approach (hope)	$(M_{\text{aff}}^-, \infty)$
M_{aff}^-	Turning point of avoidance (fear of conflict)	$(-\infty, M_{\text{aff}}^+)$
ρ_{aff}^+	Gradient of approach (hope)	$[0, \infty)$
ρ_{aff}^-	Gradient of avoidance (fear of conflict)	$[0, \infty)$
S_{aff}	Relative motivation strength	$[0, \infty)$

Once again, using a sigmoid-based model gives us additional control of the shape of the resultant tendency curve through two parameters controlling the gradients of approach and avoidance. ρ_{aff}^{+} is the gradient of hope for affiliation and ρ_{aff}^{-} is the gradient of avoidance of conflict. S_{aff} is a measure of the strength of the affiliation motivation. Again, the theory of approach-avoidance motivation [5] suggests the gradient of hope (approach) is often less than the gradient of fear (avoidance), that is, $\rho_{\text{aff}}^{+} < \rho_{\text{aff}}^{-}$. Table 2.3 summarizes the parameters of this model and the range of possible values they may take.

2.2.3 Modelling Power Motivation

Power motivation can also be modelled with respect to incentive as the difference between two sigmoid curves for tendency to seek power and inhibition of power. Tendency to seek power is lowest for low-incentive goals and highest for high-incentive goals. Negative feedback for inhibition of power is also largest for high-incentive goals. The resultant tendency for power motivation is the sum of the power-seeking and inhibition sigmoid curves, composed with incentive for success $I^{s}(G)$ as shown in Eq. 2.8. A visualization of this model is shown in Fig. 2.9.

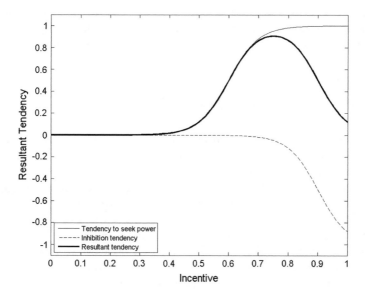

Fig. 2.9 Power motivation as the sum of curves for approaching and avoiding power ($S_{\text{pow}} = 1$, $M_{\text{pow}}^{+} = 0.6$, $M_{\text{pow}}^{-} = 0.9$ and $\rho_{\text{pow}}^{+} = \rho_{\text{pow}}^{-} = 20$). Image from [18]

Table 2.4 Parameters of the power motivation (Eq. 2.8) and their possible values

Parameter	Description	Value range
$I^s(G)$	Incentive value for success at goal G	$[0, 1]$
M_{pow}^+	Turning point of approach (power seeking)	$\left(-\infty, M_{pow}^-\right)$
M_{pow}^-	Turning point of avoidance (inhibition)	$\left(M_{\text{pow}}^+, \infty\right)$
ρ_{pow}^+	Gradient of approach (power seeking)	$[0, \infty)$
ρ_{pow}^-	Gradient of avoidance (inhibition)	$[0, \infty)$
S_{pow}	Relative motivation strength	$[0, \infty)$

$$T_{\text{pow}}^{\text{res}}(I^s(G)) = (T_{\text{pow}}^{\text{res}} \circ I^s)(G) = \frac{S_{\text{pow}}}{1 + e^{-\rho_{\text{pow}}^+\left(I^s(G) - M_{\text{pow}}^+\right)}} - \frac{S_{\text{pow}}}{1 + e^{-\rho_{\text{pow}}^-\left(I^s(G) - M_{\text{pow}}^-\right)}}$$

$$(2.8)$$

$I^s(G)$ represents incentive or the value of completing a given goal. Techniques for estimating incentive were discussed in the Sect. 2.2.2. The model has five parameters, M_{pow}^+, M_{pow}^-, ρ_{pow}^+, ρ_{pow}^- and S_{pow}. M_{pow}^+ defines the turning point of the approach component of power motivation and M_{pow}^- defines the turning point of the inhibition component of power motivation. For power motivation there is the constraint that the inhibition tendency is triggered by an increase in tendency to seek power. That is, $M_{\text{pow}}^+ < M_{\text{pow}}^-$. ρ_{pow}^+ is the gradient of tendency to seek power and ρ_{pow}^- is the gradient of inhibition. S_{pow} is a measure of the relative strength of the power motivation compared to other motives. Table 2.4 summarizes the parameters of this model and the range of possible values they may take.

2.3   Motive Profiles for Artificial Agents

When developing computational models of motivation there is a need to focus on individual motives to aid understanding of these models. However, when examining the role of motivation in goal selection there is also a need to consider several motives at once. Considering multiple motives permits the relative strengths and dominance of different motives to be taken into account when generating behaviour. The interaction of several motives changes the way an individual responds in a given situation. This section proposes three methods of differing complexity and fidelity for combining the individual motivation models above into artificial 'motive profiles'. The first is a profile of three motives, the second models only the dominant motive and the third represents motivation only in terms of the incentive value with the highest motivation.

In the sections above, affiliation and power motivation are modelled with respect to incentive, while achievement motivation is modelled with respect to probability of success. In natural systems, it is possible that the definition of incentive may change from motive to motive. However, for simplicity, in this book we assume that there is correlation (if not equality) between these definitions and further adopt Atkinson's [1] assumption that there is an inverse linear relationship between probability of success and incentive (Eq. 1.3 specifically). We can then build models as functions of a single value: incentive. This means that, if we assume we can create artificial agents that can obtain or calculate goal incentive $I^s(G)$, then we can create agents with an artificial motive profile. We do this in three ways. The first method, in Sect. 2.3.1, models profiles of three motives explicitly. The second method, in Sect. 2.3.2, models the dominant motive only. The third method, in Sect. 2.3.3, models only the incentive that maximises the dominant motive.

2.3.1 Modelling Profiles of Achievement, Affiliation and Power

We model profiles of achievement, affiliation and power motivation by combining Eqs. 2.6–2.8 as a sum, as follows:

$$
\begin{aligned}
(T^{\text{res}} \circ I^s)(G) = {}& T^{\text{res}}_{\text{ach}}(I^s(G)) + T^{\text{res}}_{\text{aff}}(I^s(G)) + T^{\text{res}}_{\text{pow}}(I^s(G)) \\
= {}& \frac{S_{\text{ach}}}{1 + e^{-\rho^+_{\text{ach}}((1-I^s(G))-M^+_{\text{ach}})}} - \frac{S_{\text{ach}}}{1 + e^{-\rho^-_{\text{ach}}((1-I^s(G))-M^-_{\text{ach}})}} \\
& + \frac{S_{\text{aff}}}{1 + e^{-\rho^+_{\text{aff}}(M^+_{\text{aff}}-I^s(G))}} - \frac{S_{\text{aff}}}{1 + e^{-\rho^-_{\text{aff}}(M^-_{\text{aff}}-I^s(G))}} \\
& + \frac{S_{\text{pow}}}{1 + e^{-\rho^+_{\text{pow}}(I^s(G)-M^+_{\text{pow}})}} - \frac{S_{\text{pow}}}{1 + e^{-\rho^-_{\text{pow}}(I^s(G)-M^-_{\text{pow}})}}
\end{aligned}
\tag{2.9}
$$

Summing the different component motives suggests that motives cooperate to influence the behaviour of an agent. Other methods of combining motives (sometimes called arbitration functions [17]) have also been proposed. For example, the combination of motives using a max(.) function models competition between motives to influence the behaviour of the agent.

This model has parameters as shown in Table 2.5. Using Eq. 2.9, we can construct artificial models of some of the named motive profiles discussed in Sects. 1.3.4 and 1.3.5. For example, a leadership motive profile [15] of high power and achievement motivation, but low affiliation motivation might appear as shown in Fig. 2.10a. An imperial motive profile of high power motivation, with low achievement and affiliation motivation might appear as shown in Fig. 2.10b.

We use Eq. 2.9 to model motivation in the experiments in Chap. 5.

Table 2.5 Parameters of motivation for an agent with a profile of achievement, affiliation and power motivation as defined in Eq. 2.9

Parameter	Description
$I^s(G)$	Incentive value for success at goal G
M_{ach}^+	Turning point of achievement approach
M_{ach}^-	Turning point of achievement avoidance
ρ_{ach}^+	Gradient of achievement approach
ρ_{ach}^-	Gradient of achievement avoidance
S_{ach}	Relative motivation strength for achievement
M_{aff}^+	Turning point of affiliation approach
M_{aff}^-	Turning point of affiliation avoidance
ρ_{aff}^+	Gradient of affiliation approach
ρ_{aff}^-	Gradient of affiliation avoidance
S_{aff}	Relative motivation strength for affiliation
M_{pow}^+	Turning point of power approach
M_{pow}^-	Turning point of power avoidance
ρ_{pow}^+	Gradient of power approach
ρ_{pow}^-	Gradient of power avoidance
S_{pow}	Relative motivation strength for power motivation

2.3.2 Modelling the Dominant Motive Only

Alternatively, we can further simplify the calculation of motivation by considering only the curve for the dominant motive. In this case we have:

$$(T^{res} \circ I^s)(G) = T_{mot}^{res}(I^s(G)) = \frac{S_{mot}}{1 + e^{-\rho_{mot}^+\left(I^s(G) - M_{mot}^+\right)}} - \frac{S_{mot}}{1 + e^{-\rho_{mot}^-\left(I^s(G) - M_{mot}^-\right)}},$$

$$(2.10)$$

where *mot* is either ach, aff or pow and parameter values are chosen from Table 2.6. We use this approach in Chap. 6.

2.3.3 Optimally Motivating Incentive

As we saw in Eqs. 2.9 and 2.10, motivational tendency for a goal is computed by composing functions for approach-avoidance motivation and incentive for success at a goal $I^s(G)$. In situations where it is desirable to avoid making the full calculation, we can approximate a motive profile by introducing the concept of an optimally motivating incentive (OMI) as follows.

First, we denote the incentive value that maximises $T^{res}(.)$ as Ω. We call Ω the OMI of the agent. Ω can be thought of as approximating the motive profile as an

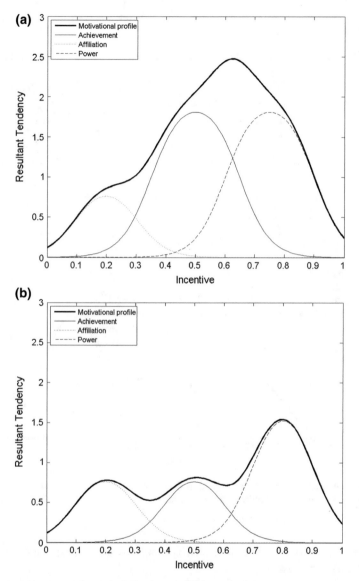

Fig. 2.10 a Leadership motive profile: $S_{ach} = 2$, $M_{ach}^+ = 0.35$, $M_{ach}^- = 0.65$, $S_{aff} = 1$, $M_{aff}^+ = 0.3$, $M_{aff}^- = 0.1$, $S_{pow} = 2$, $M_{pow}^+ = 0.6$, $M_{pow}^- = 0.9$ **b** Imperial motive profile: $S_{ach} = 1$, $M_{ach}^+ = 0.4$, $M_{ach}^- = 0.6$, $S_{aff} = 1$, $M_{aff}^+ = 0.3$, $M_{aff}^- = 0.1$, $S_{pow} = 2$, $M_{pow}^+ = 0.7$, $M_{pow}^- = 0.9$. In both profiles $\rho_{ach}^+ = \rho_{ach}^- = \rho_{aff}^+ = \rho_{aff}^- = \rho_{pow}^+ = \rho_{pow}^- = 20$. Images from [18]

agent by indicating the incentive value that will result in the highest motivation. An agent is qualitatively classified as power-motivated if its OMI is relatively 'high' in the range of possible incentives. The OMI of such an agent is shown in Fig. 2.11.

Table 2.6 Parameters of motivation and their possible values when only the dominant motive is modelled as defined in Eq. 2.10

Parameter	Description	Value range
$I^s(G)$	Incentive value for success at goal G	$[0, 1]$
M_{mot}^+	Turning point of approach	$\left(-\infty, M_{mot}^-\right)$
M_{mot}^-	Turning point of avoidance	$\left(M_{mot}^+, \infty\right)$
ρ_{mot}^+	Gradient of approach	$[0, \infty)$
ρ_{mot}^-	Gradient of avoidance	$[0, \infty)$
S_{mot}	Motivation strength	$[0, \infty)$

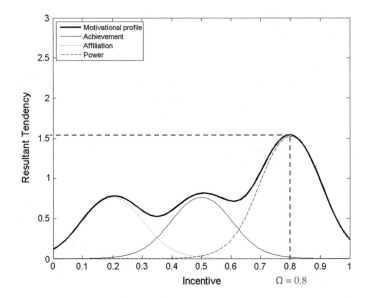

Fig. 2.11 Optimally motivating incentive Ω is the incentive value that maximises the motivation function. Power-motivated profiles such as the one shown here have a relatively 'high' optimally motivating incentive

Achievement- and affiliation-motivated agents have qualitatively 'moderate' and 'low' OMIs respectively.

Goal selection is done by computing the difference between the explicit incentive for success at a goal $I^s(G)$ and the OMI Ω and constructing a 'subjective incentive' value $\hat{I}^s(G)$ that is highest for goals with $I^s(G)$ closest to Ω. The subjective incentive value in this case is the resultant tendency for motivation. If the maximum explicit incentive value is assumed to be one, then one such equation for computing subjective incentive (resultant tendency) is:

$$(T^{\text{res}} \circ I^{\text{s}})(G) = \hat{I}^{\text{s}}(G) = 1 - |I^{\text{s}}(G) - \Omega|. \qquad (2.11)$$

That is, subjective incentive is equal to the maximum possible incentive minus the error between actual and optimal incentive. This means that a goal with an explicit incentive of one only results in the highest subjective value if it is closest to the agent's OMI Ω. The agent is assumed to prefer goals with higher subjective incentive. That is, we assume that the agent is 'subjectively rational'. The implications of this assumption are, first, that an incentive $I^{\text{s}}(G)$ will be perceived differently by agents with different OMIs. In addition, the highest explicit incentive may not be the highest subjective incentive for all agents. This provides a foundation for the emergence of behavioural diversity.

Goal selection using an OMI representation of motivation has an additional point of flexibility over the motivation functions in Sects. 2.3.1 and 2.3.2, as there is no longer a strict requirement for the range of incentive to be limited to [0, 1]. Incentive can potentially take on any range of values, including both positive (gain) and negative (loss) values. Assuming that it is possible for agents to identify the range and the maximum, V^{max}, a more general equation for subjective incentive is:

$$(T^{\text{res}} \circ I^{\text{s}})(G) = \hat{I}^{\text{s}}(G) = V^{\text{max}} - |I^{\text{s}}(G) - \Omega|. \qquad (2.12)$$

This is useful in scenarios where incentive does not conform to Atkinson's assumptions about the relationship between incentive and probability of success discussed in Chap. 1.

Yet another generalisation is possible if we weaken the definition of a goal. If the goal is merely to obtain a certain incentive with value V, then it is no longer necessary to explicitly distinguish the goal structure. This gives us the simplified equation for subjective incentive as:

$$\hat{I}^{\text{s}} = V^{\text{max}} - |V - \Omega|. \qquad (2.13)$$

This equation models the influence of implicit motivation when judging explicit incentives. The main weakness of the OMI approach is that it cannot be used to model the subtleties of hybrid motive profiles. However, for pure profiles of achievement, affiliation or power motivation, this technique can be a useful simplification. The concept of OMI is used in the motivated learning agents in Part III and the motivated evolutionary agents in Part IV.

2.4 Using Motive Profiles for Goal Selection

Goal-oriented behaviour has been widely addressed in the literature of both human psychology and artificial agents. In human psychology, goal-setting theory is considered a necessary part of motivation theories [4, 11]. Likewise, in artificial agents goals and motivations have also been closely related. Braubach et al. [3] state that the goals of an agent represent the agent's motivational stance, as it is from goals that an agent determines the actions to perform. Dignum and Conte [6] further state that truly autonomous, intelligent agents must be capable of creating new goals as well as dropping goals as conditions change. They propose instrumental goal formation as a process of deriving concrete, achievable goals—such as 'driving at the speed limit'—from high level, abstract goals—such as 'being good'. The notion of abstract goals in this case correlates somewhat with the psychological definition of implicit motives, which stem from innate preferences for certain kinds of incentives [10]. Concrete goals, in contrast, reflect explicit, self-attributed motives. There is a wealth of literature on goal structures—including goal lifecycles and type taxonomies [3]—and processes for solving goals—for machine learning, planning and rule-based agents [21].

We saw in the previous section that motivation for a goal may be computed in a number of different ways, making progressively weaker assumptions about the nature of a goal. In Eqs. 2.9–2.12 a goal is represented specifically in the equation. In Eq. 2.13 a goal is implied only by the existence of an incentive. Regardless of how the goal is represented, however, the question remains as to what to do next. That is, once motivated, how should an artificial agent choose between highly motivating goals? We discuss two traditional alternatives here: a 'winner-takes-all' approach in Sect. 2.4.1 and a probabilistic approach in Sect. 2.4.2.

2.4.1 Winner-Takes-All

In the rule-based agents studied in Part II of this book (Algorithms 3.1 and 3.2), motivation is computed using the full form of either Eq. 2.6 or Eq. 2.9. We denote by $\mathbf{G}_t = \{G^1, G^2, G^3, \ldots, G^N\}$ a set of goals that are valid at time t. We define the maximally motivating goal $G_t^{\max}$ for agent A as the element of $\mathbf{G}_t$ for which an agent A computes the highest resultant motivational tendency.

The specific calculation of $G_t^{\max}$ can be adapted to embed either a single motivation or several motivations together in a motive profile. For example, Eq. 2.14 selects a goal using achievement motivation as defined in Eq. 2.6. Equation 2.15 selects a goal using a motive profile as defined in Eq. 2.9 or an OMI as in Eq. 2.11. Chapters 3–5 examine how this motivated goal selection can be embedded in architectures for game-playing agents.

$$G_t^{\max} = \operatorname*{argmax}_{G \in \mathbf{G}_t} (T_{\text{ach}}^{\text{res}} \circ P^{\text{s}})(G) \tag{2.14}$$

$$G_t^{\max} = \operatorname*{argmax}_{G \in \mathbf{G}_t} (T^{\text{res}} \circ I^{\text{s}})(G) \tag{2.15}$$

2.4.2 Probabilistic Goal Selection

Another approach to goal selection is to select probabilistically according to the distribution of motivation values across multiple goals. This has the advantage that several goals with similar motivation values may be pursued. The probability $P(G_t = G^g)$ with which a particular goal G^g is pursued at time t may be proportional to the resultant tendency for motivation, or to the subjective incentive of the goal, or computed using a function such as the Boltzman or 'softmax' distribution to determine the probability of selecting a particular goal.

Probability proportional to the resultant tendency for motivation is computed by:

$$P(G_t = G^g) = \frac{(T^{\text{res}} \circ I^s)(G^g)}{\sum_{G \in G_t} (T^{\text{res}} \circ I^s)(G)}.$$ (2.16)

This approach is used in Chap. 6. Boltzman goal selection using a motive profile such as Eq. 2.10 or Eq. 2.11 is computed by:

$$P(G_t = G^g) = \underset{G \in G_t}{\text{softmax}}(T^{\text{res}} \circ I^s)(G) = \frac{e^{\frac{(T^{\text{res}} \circ I^s)(G^g)}{\tau}}}{\sum_{G \in G_t} e^{\frac{(T^{\text{res}} \circ I^s)(G)}{\tau}}}.$$ (2.17)

$0 < \tau < \infty$ is a temperature value that determines the difference in probability of goals with high motivation values from goals with low motivation values. τ can be varied to increase or decrease the probability of the agent executing a randomly selected goal, or it can be kept constant.

2.5 Summary

This chapter has presented three incentive-based computational models of motivation for achievement, affiliation and power motivation. The models use the concept of approach and avoidance motivation and include curves for both approach and avoidance of a particular motivation. Control parameters permit the maximally motivating incentive to be modified in each model, as well as the rate of increase and decrease of motivation as subjective probability of success or goal incentive changes.

The models are designed such that they can be used in isolation or together, embedded in an artificial 'motive profile'. A motive profile can be further approximated as an optimally motivating incentive.

We have provided a formal notation for selection of a maximally motivating goal from a set of goals, or probabilistic goal selection. However, we have not yet addressed the questions of how goals are created, or where success probabilities or incentive values come from. Chapter 3 will describe how computational models of motivation can be combined with some traditional agent architectures for game-playing agents.

References

1. J.W. Atkinson, Motivational determinants of risk-taking behavior. Psychol. Rev. **64**, 359–372 (1957)
2. J.W. Atkinson, G.H. Litwin, Achievement motive and test anxiety conceived as motive to approach success and motive to avoid failure. J. Abnorm. Soc. Psychol. **60**, 52–63 (1960)
3. L. Braubach, A. Pokahr, D. Moldt, W. Lamersdorf, Goal representation for BDI agent systems, in *Proceedings of the Second International Workshop on Programming Multiagent Systems: Languages and Tools*, 2005, pp. 9–20 (2005)
4. N. Brody, *Human Motivation: Commentary on Goal-directed Action* (Academic Press Inc., London, 1983)
5. J.S. Brown, Gradients of approach and avoidance responses and their relation to motivation. J. Comp. Physiol. Psychol. **41**, 450–465 (1948)
6. F. Dignum, R. Conte, Intentional agents and goal formation, in *Proceedings of the Intelligent Agents IV: Agent Theories, Architectures and Languages*, 1998, pp. 231–243
7. A. Elliot, *Handbook of Approach and Avoidance Motivation* (Taylor and Francis, New York, NY, 2008)
8. A. Elliot, A. Eder, E. Harmon-Jones, Approach-avoidance motivation and emotion: convergence and divergence. Emot. Rev. **5**, 308–311 (2013)
9. P. Galanter, Computational aesthetic evaluation: past and future, in *Computers and Creativity*, ed. by J. McCormack, M. d'Inverno (Springer, Berlin, 2012), pp. 255–293
10. J. Heckhausen, H. Heckhausen, *Motivation and Action* (Cambridge University Press, New York, NY, 2010)
11. E. Locke, G. Latham, *A Theory of Goal Setting and Task Performance* (Prentice Hall, New Jersey, USA, 1990)
12. B.A. Maher, The application of the approach-avoidance conflict model to social behavior. J. Conflict Resolut. **8**, 287–291 (1964)
13. J. McClelland, J.W. Atkinson, R.A. Clark, E.L. Lowell, A scoring manual for the achievement motive, in *Motives in Fantasy, Action and Society*, ed. by J.W. Atkinson (D. Van Nostrand Company, Inc., Princeton, NJ, 1958)
14. J. McClelland, R.I. Watson, Power motivation and risk-taking behaviour. J. Pers. **41**, 121–139 (1973)
15. J. McClelland, *Power: The Inner Experience* (Irvington, New York, 1975)
16. J. McCormack, M. d'Inverno (eds.), *Computers and Creativity* (Springer, Berlin, 2012)
17. K. Merrick, M.L. Maher, *Motivated Reinforcement Learning: Curious Characters for Multiuser Games* (Springer, Berlin, 2009)
18. K. Merrick, K. Shafi, Achievement, affiliation and power: motive profiles for artificial agents. Adapt. Behav. **19**, 40–62 (2011)
19. N.E. Miller, Experimental studies of conflict, in *Personality and the behavior disorder*, ed. by J.M. Hunt (Ronald, New York, NY, 1944)
20. J. Nikitin, A. Freund, When wanting and fearing go together: the effect of co-occurring social approach and avoidance motivation on behavior, affect and cognition. Eur. J. Soc. Psychol. **40**, 783–804 (2009)

21. S.J. Russell, P. Norvig, *Artificial Intelligence a Modern Approach* (Prentice Hall, New Jersey, USA, 2010)
22. R. Saunders, J.S. Gero, Curious agents and situated design evaluations. Artif. Intell. Eng. Des. Anal. Manuf. **18**, 153–161 (2004)

Chapter 3
Game-Playing Agents and Non-player Characters

This chapter brings together the background and theory from the previous two chapters in four agent architectures for game-playing agents. It describes how motivation can be embedded in rule-based agents, crowds, learning agents and evolutionary algorithms. Motivated rule-based agents are suitable for decision making by individual agents, while motivated learning agents are suitable for competitive or strategic decision making when two or more agents interact. The motivated crowd and evolutionary algorithms are suitable for controlling groups of agents. These architectures are the topics of study in Part II, Part III and Part IV of the book.

3.1 Artificial Intelligence in Non-player Characters

In a recent paper, Yannakakis [33] identifies four flagship research areas related to artificial intelligence in games. These are player experience modelling, game data mining, procedural content generation, and non-player character artificial intelligence.

As the size and demographic diversity of virtual world users (including of computer games) increases, there is a need to tailor virtual environments to support the needs of users with different skills and experiences. Player experience modelling describes the use of computational techniques for constructing computational models of a player's experiences and satisfaction. Related to player experience modelling, game data mining seeks a better understanding of how and why players play games. The work by Yee [34] discussed in Chap. 1 is an example of this relatively new field of research.

Procedural content generation is the process of generating game content automatically. Game content includes elements such as terrains, maps, levels, stories, quests and music, but generally excludes non-player character behaviour, which is considered an independent research challenge. Non-player characters are computer-controlled characters that can be enemies, opponents, pets or partners to

© Springer International Publishing AG 2016   45
K.E. Merrick, *Computational Models of Motivation for Game-Playing Agents*,
DOI 10.1007/978-3-319-33459-2_3

human-controlled player characters. Alternatively, they may simply support the plot of the game. Artificial intelligence research has provided solutions for tasks such as navigation and control and the focus is now turning to new research areas that can increase the capabilities of non-player characters. These include generation and detection of patterns of complex social behaviour and interaction among non-player characters and humans.

Broadly speaking, the first two research areas—player experience modelling and game data-mining—are concerned with understanding game players. The second two are concerned with building more adaptable, believable and exciting virtual worlds. However, the topics are interrelated. For example, there is potential for player experience models to inform the procedural generation of game content that is tailored to the specific experiences of a given player. In this book we use information gathered by game data mining researchers about players to inform the design of game-playing agents that can be used to create non-player characters with novel and diverse decision-making characteristics.

We adapt four existing technologies used for non-player characters to embed motivation: rule-based agents, a crowd algorithm, learning agents and an evolutionary algorithm. We describe these adaptations in the remaining sections of this chapter. Section 3.2 considers motivated rule-based agents and the assumptions we make about the scenarios to which such agents might be suited. Section 3.3 considers crowds of motivated agents. Section 3.4 presents an algorithm for motivated learning agents and Sect. 3.5 presents an algorithm for the evolution of a society of motivated agents. Each section includes an algorithm defining the intelligence loop of the corresponding agent. These algorithms are then analysed and applied in Parts II, III and IV of this book.

3.2   Rule-Based Agents

Rule-based agents are embedded with a set of rules about states of the game world of the form: `if <condition> then <behaviour>`. If a non-player character (NPC) controlled by a rule-based agent observes a state that fulfils the `<condition>` of a rule, then the corresponding `<behaviour>` is executed. Only states of the world that meet a `<condition>` will produce a `<behaviour>` in response.

Rule-based agents essentially have preset behaviours. They thus represent a fairly basic form of intelligent agent. However, with a variety of rules and actions, the overall result can be a behaviour system that is not obvious to human players [18]. Kehoe [18] describes one example of a rule-based agent controlling a dealer character in the card game *Blackjack*. The dealer has a simple rule as follows: If the cards add up to 17 or less, then 'hit' (take an extra card). To the average player, the perception is that the dealer is playing competitively. Unless the house advertises the rule used by the dealer, the player will most likely believe the adversary is more competent than the one he or she actually faces.

Another classic application of rule-based agents is to control the ghosts in the iconic 1980s arcade game *Pac-Man* [19]. Four ghosts chase a player through a maze. Each ghost follows a slightly different rule set to introduce diversity and personality into the NPCs. The red ghost always chases the player. The pink ghost selects random moves and acts fast. The orange ghost also selects random moves, but acts slowly. The blue ghost initially avoids the player. However, if the player approaches the blue ghost then it will chase the player. The diversity of the *Pac-Man* ghosts is considered central to the success of the game [19]. If all the ghosts simply chased the player, they would line up behind the player and three of them would become irrelevant. If they were too fast, the game would be too hard. Too slow and the game would be too easy. In games where there are thousands, rather than a handful, of NPCs, however, implementing such diversity becomes a significant challenge.

In *Pac-Man* and other traditional rule-based agents, the rules that define personality are strongly tied to the game logic. The approaches in this book, instead, base rules on a generic set of motivational variables, so that the agents themselves can execute their decision-making processes for any given set of goals and incentives. The goals and situational characteristics from which incentive is calculated may be domain-specific, but a given agent can have a domain-independent set of motivations. The next section considers the types of game scenarios that would be well suited to such agents, and Sect. 3.2.2 presents an algorithm for motivated rule-based agents. Such agents are the topic of the experiments in Chaps. 4 and 5, as well as the first half of Chap. 6.

3.2.1 Game Scenarios for Motivated Rule-Based Agents

As we saw in Chaps. 1 and 2, motivation plays a significant role in influencing people's decision making when they are faced with different strength incentives. The if-then rules in traditional rule-based agents essentially map conditions about states to behaviours. In motivated agents we introduce an intermediate construct in the form of a goal. A condition about a state is mapped to a set of goals that are 'valid' in that state. Each goal is assumed to have a different incentive. The agent must select one goal from the set of valid goals and execute its corresponding behaviour to solve the goal. Agents with different motivational preferences will select different goals in any given state. This is the basis for the emergence of behavioural diversity.

Sets of goals with different incentives occur naturally in many game scenarios. For example, a character might have to estimate how close to an enemy to creep before firing a ranged weapon. Each firing position represents a different goal with a different probability of success and thus a different incentive. The ring-toss game studied in Chap. 4 has properties similar to this. Another character might be faced with several quests for different treasures, each guarded by a different opponent. The opponents may have different physical characteristics that influence the perceived probability of success of an attack, while the treasures may have different values. Both of these factors thus influence the incentive for each quest. The roulette game studied in Chap. 5 has these characteristic properties.

The next section introduces two motivated rule-based agent algorithms. The first uses winner-takes-all goal selection. We use this model in Chaps. 4 and 5 to examine the decision-making characteristics of different models of motivation. The second algorithm uses probabilistic goal selection. We will use this algorithm in Chap. 6.

3.2.2   Motivated Rule-Based Agents

As outlined above, condition-behaviour rules in traditional rule-based agents are replaced by condition-goal-behaviour rules in motivated rule-based agents. Multiple goals may be valid under a given condition, but motivation determines which goal fires. Motivation is thus a 'meta-condition' in a motivated rule-based agent. Because it is possible for an agent's motivation to change over time, the rule that fires in response to a particular condition may also change over time.

Algorithm 3.1 shows the intelligence loop for a motivated rule-based agent. The agent is assumed to work with a map of condition-goal-behaviour tuples (rules). A behaviour B is a sequence of actions that fulfils the goal to which it is mapped. For example, a behaviour that fulfils the goal of killing an enemy might include actions to move towards the enemy, draw a weapon and use the weapon. In the demonstrations in Chaps. 4 and 5 we assume the existence of a master list of tuples from which subsets are constructed. However, in Chap. 8 we present an original game where the list of tuples is filled on the fly by the player, and the agents adapt their behaviour to the available goals.

The algorithm proceeds as follows: At each time t, the agent senses a state $S(W_t)$ (line 3). The agent constructs a subset of goals $\mathbf{G}_t$ that are valid in the current state $S(W_t)$ (line 4). From these goals the agent computes a maximally motivating goal $G_t^{\max}$ by applying a computational model of motivation such as Eq. 2.14 or 2.15 (line 5). This is winner-takes-all goal selection. Finally, the appropriate behavioural response to that goal is executed (line 6). The entire intelligence loop (lines 3–6) is then repeated as long as the agent is alive.

Algorithm 3.2 shows the intelligence loop for a variant of the motivated rule-based agent that uses probabilistic goal selection. The algorithm follows the same steps as Algorithm 3.1, but line 5 in Algorithm 3.1 is replaced by probabilistic goal selection in Algorithm 3.2. This means that the most highly motivating goal is selected with the highest probability, but other less motivating goals may also be selected with some probability proportional to their motivation.

In the next section we extend the structure of the motivated rule-based agents in Algorithm 3.1 to model crowds of motivated agents.

Algorithm   3.1 (*A   motivated   rule-based   agent   with   winner-takes-all   goal selection*)

1.	Initialise motivation parameters in Table 2.5, 2.6 or an OMI
2.	Repeat for each time t:
3.	Sense the current state of the world $S(W_t)$
4.	Construct a goal set $\mathbf{G}_t = \{G^g \mid \exists \langle C, G^g, B \rangle$ where $S(W_t)$ satisfies $C\}$
5.	Compute G_t^{max} using Equation 2.14 or 2.15
6.	Execute behaviour B mapped to G_t^{max}

Algorithm 3.2 (*A motivated rule-based agent with probabilistic goal selection*)

1.	Initialise motivation parameters in Table 2.5, 2.6 or an OMI
2.	Repeat for each time t:
3.	Sense the current state of the world $S(W_t)$
4.	Construct a goal set $\mathbf{G}_t = \{G^g \mid \exists \langle C, G^g, B \rangle$ where $S(W_t)$ satisfies $C\}$
5.	Compute $(T^{res} \circ I^s)(G)$ for all G using Equation 2.9, 2.10, 2.11 or 2.12
6.	Select a goal G_t^g using Equation 2.16 or 2.17
7.	Execute behaviour B mapped to G_t^g

3.3   Crowds

Sometimes it is appropriate for NPCs to move in cohesive groups, rather than as individuals [28]. Birds, animals such as sheep and cattle, and fish are everyday examples of creatures we might expect in flocks, herds or schools. In games we might also see herds of monsters, non-player combat units might be controlled by flocking behaviour, or a group of support characters might exhibit characteristics of life-like crowd behaviour.

At the heart of computational models of flocks, herds, schools, swarms and crowd behaviour is Reynold's iconic *boids* model [26]. The *boids* model can be viewed as a kind of rule-based reasoning in which rules take into account properties of other agents. The three fundamental rules are:

- *Cohesion*: Each agent moves toward the average position of its neighbours;

- *Alignment*: Each agent steers so as to align itself with the average heading of its neighbours;

- *Separation*: Agents move to avoid hitting their neighbours.

In *boids*, behaviours are implemented as forces that act on agents when a certain condition holds. Suppose we have a group of n agents $A^1, A^2, A^3, \ldots, A^n$. At time t

each agent A^j has a position $\mathbf{p}_t^j$, a direction $\hat{\mathbf{d}}_t^j$ and a speed S. $\mathbf{p}_t^j$ is a point and $\hat{\mathbf{d}}_t^j$ a unit vector. At each time step t, the direction of each agent is updated as follows:

$$\vec{\mathbf{d}}_{t+1}^j = W_d \hat{\mathbf{d}}_t^j + W_c \hat{\mathbf{c}}_t^j + W_a \hat{\mathbf{a}}_t^j + W_s \hat{\mathbf{s}}_t^j. \tag{3.1}$$

$\vec{\mathbf{d}}_{t+1}^j$ is then normalised (each component d_x^j, d_y^j etc. is divided by the magnitude of the vector to give a new direction $\hat{\mathbf{d}}_{t+1}^j$). $\hat{\mathbf{c}}_t^n$ is a unit vector in the direction of the average position of agents within a certain range of A^j (called the neighbours of A^j); $\hat{\mathbf{a}}_t^j$ is a unit vector in the average direction of agents within a certain range of A^j; and $\hat{\mathbf{s}}_t^j$ is a unit vector in the direction away from of the average position of agents within a certain range of A^j. These vectors are the result of cohesive, alignment and separation forces corresponding to the rules outlined above. Weights W_c, W_a and W_s strengthen or weaken the corresponding force. W_d strengthens or weakens the perceived importance of the original direction of travel. Once a new direction of travel has been computed, the position of each agent is updated by:

$$\mathbf{p}_{t+1}^j = \mathbf{p}_t^j + S\hat{\mathbf{d}}_{t+1}^j. \tag{3.2}$$

As we noted above, agents that are within a certain range R of a particular agent A^j are called its neighbours. Formally, we can define the subset $\mathbf{N}^j$ of agents within a certain range R of A^j as follows:

$$\mathbf{N}^j = \left\{ A^k | A^k \neq A^j \wedge dist\left(A^k, A^j\right) < R \right\}, \tag{3.3}$$

where $dist\left(A^k, A^j\right)$ is generally the Euclidean distance between two agents.

Different ranges may be used to calculate cohesive, alignment and separation forces. In this case they are denoted by R_c, R_a and R_s. The average position $\mathbf{c}_t^j$ of agents within range R_c of A^j (excluding A^j) is calculated as:

$$\mathbf{c}_t^j = \frac{\sum_k \mathbf{p}_t^k}{\left| (\mathbf{N}_c^j)_t \right|}. \tag{3.4}$$

The vector in the direction of this average position is calculated as:

$$\vec{\mathbf{c}}_t^j = \mathbf{c}_t^j - \mathbf{p}_t^j. \tag{3.5}$$

$\vec{\mathbf{c}}_t^j$ is normalised to get the unit vector $\hat{\mathbf{c}}_t^j$ in the same direction. Similarly, we can calculate the average position $\mathbf{s}_t^j$ of agents within range R_s of A^j (excluding A^j) as

$$\mathbf{s}_t^j = \frac{\sum_k \mathbf{p}_t^k}{|(\mathbf{N}_s^j)_t|}. \tag{3.6}$$

The unit vector $\hat{\mathbf{s}}_t^j$ away from this position is calculated by normalising the difference:

$$\vec{\mathbf{s}}_t^j = \mathbf{p}_t^j - \mathbf{s}_t^j. \tag{3.7}$$

Finally, the unit vector $\hat{\mathbf{a}}_t^j$ in the average direction of agents within range R_a of A^j (excluding A^j), is calculated by normalising the sum:

$$\vec{\mathbf{a}}_t^j = \frac{\sum_k \hat{\mathbf{d}}_t^k}{|(\mathbf{N}_a^j)_t|}. \tag{3.8}$$

The basic *boid* flocking algorithm does not incorporate mechanisms for goal-directed behaviour or obstacle avoidance. However, these can be modelled as additional forces with relative ease. We do this in the demonstrations in Chap. 8. The next section presents an algorithm for crowds of motivated agents in which motivation is implemented as forces that pull agents towards goals depending on the incentive of the goal and the motive profile of the agent.

3.3.1 Motivated Crowds

Some previous work has considered the use of motivation in crowds, including Saunders' curious design agents [27]. Algorithm 3.3 takes a similar approach, modelling motivation as rules for the application of additional forces.

The algorithm proceeds as follows: Each agent in the crowd is initialised with a model of motivation (line 1). At each time step, each agent senses the local state of its environment (line 4), including the features described above, so $S(W_t) = \langle \mathbf{p}_t^j, \hat{\mathbf{d}}_t^j, (\mathbf{N}_c^j)_t, (\mathbf{N}_s^j)_t, (\mathbf{N}_a^j)_t, S \rangle$. Each agent then constructs a set of goals that conform to a condition on the current state (line 5). For crowd agents, the condition might concern proximity to a goal. For example, the goal set might contain goals within range R_m of the agent's current position, and with a sufficiently large subjective incentive. That is, C asserts that:

$$dist(\mathbf{g}_t^g, \mathbf{p}_t^j) < R_m \wedge (T^{res} \circ I^s)(G^g) > \gamma, \tag{3.9}$$

Algorithm 3.3 (*A crowd of motivated agents*)

1.   Initialise n and a society $\mathbf{A}$ of n agents with position, velocity, weights, ranges and motivation parameters in Table 2.5, 2.6 or an OMI
2.   Repeat for each time t:
3.   Repeat for each agent A^j
4.   Sense the current local state $S(W_t) = \langle \mathbf{p}_t^j, \hat{\mathbf{d}}_t^j, (\mathbf{N}_c^j)_t, (\mathbf{N}_s^j)_t, (\mathbf{N}_a^j)_t, S \rangle$
5.   Construct goal set $\mathbf{G}_t = \{G^g \mid \exists \langle C, G^g, B \rangle$ where $S(W_t)$ satisfies $C\}$
6.   Compute $\hat{\mathbf{g}}_t^g$ using Equation 3.10 for all g
7.   Sum all forces on agent A^j using Equation 3.11
8.   Move all agents to new positions using Equation 3.2.

where $\mathbf{g}_t^g$ is the position of the goal G^g. A force in the direction of $\mathbf{g}_t^g$ is then calculated (lines 6) as:

$$\vec{\mathbf{g}}_t^g = \mathbf{g}_t^g - \mathbf{p}_t^j. \tag{3.10}$$

$\vec{\mathbf{g}}_t^j$ is normalised to get the unit vector $\hat{\mathbf{g}}_t^j$ in the same direction. This force is included in the update equation for the agent (line 7):

$$\vec{\mathbf{d}}_{t+1}^j = W_d \hat{\mathbf{d}}_t^j + W_c \hat{\mathbf{c}}_t^j + W_a \hat{\mathbf{a}}_t^j + W_s \hat{\mathbf{s}}_t^j + W_m \sum_g \hat{\mathbf{g}}_t^g. \tag{3.11}$$

Finally all agents are moved to their new positions (line 8).

This algorithm represents an extension of the rule-based approaches in Sect. 3.2. It takes motivation from a single agent to a multi-agent setting, with specific definitions for the sensed state, conditions and behaviours. However, the strategies employed by individual agents remain static over time. In the next section we return to the single agent setting to consider the role of learning in motivated agents.

3.4   Learning Agents

Unlike rule-based agents, learning agents are able to modify their internal structure in order to improve their performance with respect to some task [24]. Learning agents in computer games are still relatively rare, due to the computational resources required for learning, and a fear that permitting computer-controlled characters to learn will result in unpredictable game play. The approach taken to learning in this book aims to address this issue by defining the learning scenarios using game-theoretic models. These scenarios can then be analysed for properties such as equilibrium points so that game designers can better predict the possible learning outcomes in each scenario.

An early example of a game in which learning agents control NPCs is *Black and White*. In *Black and White*, NPCs (specifically pets) can be trained to learn behaviours specified by their human master. The human provides the NPC with a reward such as food or a pat to encourage desirable actions and punishment to discourage unwanted actions. In the more recent *Creatures* [11], NPCs called *Norns* are motivated by biological drives, have artificial neural networks for learning and can be cross-bred to evolve new kinds of *Norns*. *Creatures* is considered significant for being one of the first commercial games to include complex artificial life algorithms.

Drivatars (short for driving avatars) [23] in the *Microsoft Xbox* game *Forza Motorsport* are artificial characters that can learn from training examples provided by a human. Researchers from *Microsoft* [9] have also shown that it is possible to use reinforcement learning (RL) to allow NPCs to develop new skills by applying RL to fighting characters for the *Xbox* game *Tao Feng*.

Other research frameworks have also considered motivation in a learning context, including opportunistic motivated learning (OML) [10] and motivated reinforcement learning (MRL) [21]. MRL agents, for example, first identify which skills they would like to learn and then learn behaviours to perform the chosen skills. In contrast to MRL agents, the motivated learning agents presented in this book are concerned with learning high-level sequences of behaviours (called strategies) that satisfy their motives, rather than low-level action sequences that comprise a behaviour. The focus of this book is on learning during strategic interactions modelled by two-player social dilemma games. We thus take some time in Sect. 3.4.1 to define the structure of such games.

3.4.1 Social Dilemma Games

In a two-player social dilemma game, each of two players has two choices: to cooperate with the other player—denoted by B^C—or defect (refuse to cooperate), denoted by B^D. Once both players have made a choice and executed their chosen behaviour, each receives a payoff (gain or loss) V. The payoff might be money, points in the game, an injury or damage to equipment. The payoff for each player depends on the choices made by both players. This pair of choices is called a game outcome. An outcome is represented by a bracketed pair. For example, (B^D, B^D) is the outcome when both players chose B^D. Table 3.1 summarises the payoff

Table 3.1 The payoff structure of a two-player social dilemma game

		Player 2	
Player 1		B^D	B^C
	B^D	P, P	T, S
	B^C	S, T	R, R

R denotes the reward for mutual cooperation, P the punishment for mutual defection, T the temptation to defect and S the sucker's payoff

structure of a two-player social dilemma game. Rows in Table 3.1 represent the choices available to Player 1. Columns represent the choices available to Player 2. For a given outcome, the payoff for Player 1 is the first value in the cell corresponding to the game outcome. The payoff for Player 2 is the second value in the same cell. Table 3.1 shows that the different outcomes result in different payoffs to the players: R, T, S and P. These values have specific meanings as follows. R denotes the 'reward' for mutual cooperation. P denotes the 'punishment' for mutual defection. T represents the 'temptation' to defect from the (B^C, B^C) outcome, while S is the sucker's payoff for choosing B^C when the other player chooses B^D. Inspection of Table 3.1 shows that the payoff structure is the same for both players.

In a more formal notation, the game $\mathbf{W}$ presents a player with a matrix:

$$\mathbf{W} = \begin{bmatrix} P & T \\ S & R \end{bmatrix}. \tag{3.12}$$

This notation permits us to manipulate games mathematically or computationally.

Abstract social dilemma games can be used to model in-game scenarios or mini-games that might occur within a larger-scale computer game. The seemingly simple game above, for example, is in fact powerful enough to represent a range of scenarios that occur when individuals interact in their everyday lives and in virtual worlds. For example, if we assume $T > R > P > S$, then the game can model an arms race, such as commonly occurs in turn-based strategy (TBS) games like Sid Meier's *Civilisation* [20] series. In the *Civilisation* series games, NPCs have a choice of pursuing a peaceful, scientific victory, or a war-like military victory. The B^D choice represents assignment of resources to build weapons and the B^C choice represents assignment of resources to other national priorities. A NPC that chooses B^D when the other player does not has an advantage because they possess weapons that could potentially subdue the other. When both parties build weapons (the (B^D, B^D) outcome), there is no such advantage, but both parties have put their resources into the development of weapons at the expense of other national priorities. The (B^C, B^C) outcome thus has a higher payoff as both parties can assign their resources elsewhere to increase the size of their cities and the happiness of their subjects. We see from this example that B^C and B^D are high-level behaviours, not individual actions. For example, B^D might comprise a series of actions for building military units, fortifying cities and so on. B_t is used to denote the behaviour chosen at decision-making time t, with the understanding that a series of game-specific actions may then be executed before another decision is made at time $t + 1$.

We now have three important concepts: strategies, behaviours and actions. A strategy is a plan of play that will satisfy the agent's motives. In other words, a strategy is a sequence of behaviours B. In general, a series of actions might then be required to execute each behaviour. We assume the existence of behaviours encapsulating such actions and concern ourselves with the task of learning the plan of play. The next section formalises this idea of a strategy as a plan of play.

3.4.2 Strategies in Game Theory

Formally, a strategy σ is a function that takes a game as input and outputs a decision B_t. A strategy may be pure, such as 'always choose B^C', or mixed. Mixed strategies make a stochastic choice between two pure strategies with a fixed frequency. Suppose we denote the probability that Player 2 will choose B^C as $P^2(B_t = B^C)$; then the expected payoff for the two pure strategies available to Player 1 ('always play B^C' or 'always play B^D') can be computed as follows:

$$E^1(B^C) = P^2(B_t = B^C)R + [1 - P^2(B_t = B^C)]S, \tag{3.13}$$

$$E^1(B^D) = P^2(B_t = B^C)T + [1 - P^2(B_t = B^C)]P. \tag{3.14}$$

Using this information, a player can identify the strategy with the maximum expected payoff. A variation on this idea that takes into account individual differences in preference is utility theory [17]. Utility theory acknowledges that the values of different outcomes for different people are not necessarily equivalent to their raw payoff values. If the payoff value is denoted by V, a utility function $U(V)$ is a twice differentiable function defined for $V > 0$, which has the properties of non-satiation (the first derivative $U'(V) > 0$) and risk aversion (the second derivative $U''(V) < 0$). The non-satiation property implies that the utility function is monotonic, while the risk aversion property implies that it is concave. Utility theories were first proposed in the 1700s and have been developed and critiqued in a range of fields including economics [16] and game theory [31].

Alternatives have also been proposed to model effects that are inconsistent with utility theory. Examples include prospect theory [16] and lexicographic preferences [6]. The models in this book can also be thought of as an alternative to utility theory that uses theories of motivation to determine how to compute individuals' preferences.

Various other techniques have been proposed to model decision making under uncertainty, that is, when it is not possible to assign meaningful probabilities to alternative outcomes. Many of these techniques capture 'rules of thumb' or heuristics used in human decision making [4, 7]. Examples include the maximax, maximin and regret principles.

The strategies chosen by players and their corresponding payoffs constitute a Nash Equilibrium (NE) if no player can benefit by changing their strategy while the other player keeps theirs unchanged. This definition covers mixed strategies in which players make probabilistic random choices. Formally, if we consider a pair of strategies, σ^1 and σ^2, and denote the expected payoff for Player 1 using σ^1 against Player 2 using σ^2 as $E^1(\sigma^1, \sigma^2)$, then the two strategies are in equilibrium if $E^1(\sigma^1, \sigma^2) \geq E^1(\bar{\sigma}^1, \sigma^2)$ for all $\bar{\sigma}^1 \neq \sigma^1$. In other words, the strategies are in equilibrium if there is no alternative strategy for Player 1 that would improve his expected payoff against Player 2 if Player 2 continues to use strategy σ^2 [12].

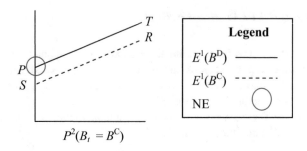

Fig. 3.1 Visualization of the payoff expected by Player 1 in a social dilemma game with $T > R > P > S$. The expected payoff for two different behaviours is plotted against the probability of Player 2 choosing the behaviour B^C. The Nash Equilibrium (NE) is circled

In the arms race game described above, there is a pure strategy equilibrium point (B^D, B^D) from which neither player benefits from unilateral deviation, although both benefit from joint deviation. We can visualize this game in terms of expected payoff as shown in Fig. 3.1. We denote the probability of Player 2 choosing B^C as $P^2(B_t = B^C)$, the expected payoff if Player 1 chooses B^D as $E^1(B^D)$, and the expected payoff for Player 1 choosing B^C as $E^1(B^C)$. The visualization shows that assuming $T > R > P > S$ implies $E^1(B^D) > E^1(B^C)$ regardless of $P^2(B_t = B^C)$. In other words, the strategy of choosing B^D dominates the strategy of choosing B^C.

For this reason, classical game theory predicts that play will ultimately converge on the (B^D, B^D) outcome when playing this particular social dilemma game. However, in practice that is generally not the case. Humans with different motives consider alternative strategies that avoid the (B^D, B^D) outcome. Likewise, in virtual worlds, NPCs are programmed with rules to control when they offer a peace treaty and when they declare war. These rules might, for example, be based on the military strength of the opposing player. In this book, we consider an alternative approach that embeds game-playing agents with computational motivations. The advantage of this approach is that these motives can be used in any scenario that can be modelled by a game of the form in Table 3.1. The domain-specific details of the scenario no longer need be considered. In addition, game theory provides us with techniques for predicting learning trajectories of agents, so there is reduced uncertainty about the outcome of learning.

3.4.3   Motivated Learning Agents

In developing the motivated learning agent algorithm in this section, we make four assumptions [22]:

- Payoff is used to represent incentive;

- Individual agents are subjectively rational in their quest for accumulating this payoff;
- There is no communication between opposing agents to give warning of, or otherwise negotiate, the behaviours each will select;

- The motivations of other agents are not observable.

The first assumption means that we are focusing on the influence of implicit motives when judging explicit incentives. The second assumption means that different agents may perceive the same explicit incentive (payoff) differently as a result of having different motives. This assumption, in fact, underlies all the work in this book. Other studies have considered the case where individuals are subjectively irrational [2, 5]. In our case, the assumption of subjective rationality hinges on the link between motivation and action. In particular, extensive experimental evidence indicates that individuals will act, apparently sub-optimally according to some objective function, to fulfil their (subjective) motives [15].

Non-communication between agents is a standard game-theoretic assumption and we adopt it in this book. By non-communication, here we mean non-communication of the choice of behaviour before it is made. Once the choice is made, it may be communicated to or otherwise perceivable by the other agent. For example, once an agent has decided to pursue a scientific path to victory in a civilisation game, this decision may be revealed through an offer of a peace treaty to the opponent.

The non-observability of motivations is assumed based on the difficulty of identifying an individual's motive profile, particularly during short, casual interactions. This assumption differentiates the work in this book from other work on altered perception or misperception where the perceptions of others are assumed to be observable [2]. Specifically, we do not model perception of a game in terms of expectations about what other players may do. Such approaches have been taken in other related game [5], metagame [4] or hypergame [32] theory research concerning misperception or the evolution of preferences. Rather, we model perception as a process of transforming one game into another.

As we have seen, in a two-player social dilemma game, there are four possible outcomes: (B^C, B^C), (B^D, B^D), (B^C, B^D) and (B^D, B^C). We can think of these outcomes as implying four goals, G^R, G^P, G^S and G^T. However, as these goals are merely to obtain a certain payoff ($V = R$, $V = P$, $V = S$ or $V = T$ respectively), we make the simplifying assumption that $I^s(G) = V$ as described in Sect. 2.3.3.

In our motivated learning agent algorithm (Algorithm 3.4), the agent is initialised with an OMI Ω and initial probabilities $P(B_0 = B^C)$ and $P(B_0 = B^D)$. $P(B_0 = B^C)$ is the probability that the behaviour B^C is executed at time $t = 0$. $P(B_0 = B^C)$ is chosen at random from a uniform distribution, and $P(B_0 = B^D) = 1 - P(B_0 = B^C)$.

When $t = 0$, the agent selects an action probabilistically based on $P(B_0 = B^C)$ and $P(B_0 = B^D)$ as shown in lines 8–9. In each subsequent iteration, the agent receives a payoff V_t for its previously executed behaviour (line 5) and then computes a subjective incentive value informed by its individual motive profile (line 6).

Using the methodology in Sect. 2.3.3, we assume that each motivated learning agent wishes to adopt a strategy that results in an outcome that minimizes the difference between V_t and its OMI Ω. We use Eq. 2.13 for this.

The agent also recalls the last behaviour selected (line 5) and uses all this information to update its probabilities of choosing B^C and B^D (line 7). Cross learning formalised for two-player games [3] is used to model this cultural learning as follows:

$$P(B_t = B^i) = \begin{cases} [1 - \alpha \hat{I}_t^s]P(B_{t-1} = B^i) + \alpha \hat{I}_t^s & \text{if } B_{t-1} = B^i \\ [1 - \alpha \hat{I}_t^s]P(B_{t-1} = B^i) & \text{otherwise} \end{cases}, \qquad (3.15)$$

where i is either C or D.

The next behaviour is then chosen probabilistically based on these updated values (line 9). The chosen behaviour is then executed (line 10). The intelligence loop in lines 4–10 is repeated as long as the agent is alive.

Algorithm 3.4 (*A motivated learning agent*)

1.   Initialise game world **W** of the form in Equation 3.12 and identify V^{max}
2.   Initialise optimally motivating incentive Ω and $P(B_0 = B^C)$ and $P_0(B_0 = B^D)$
3.   Repeat for each time t:
4.   if $t > 0$
5.   Receive payoff V_t for previously executed behaviour B_{t-1}
6.   Compute subjective incentive $\hat{I}_t^s$ using Equation 2.13
7.   Compute $P(B_t = B^C)$ and $P(B_t = B^D)$ using Equation 3.15
8.   Generate a random number r from $[0, 1]$
9.   if $r < P(B_t = B^C)$ select behaviour $B_t = B^C$ else select behaviour $B_t = B^D$
10.   Store and execute B_t

Part III examines the use of Algorithm 3.4 in a number of scenarios that might arise in the context of a larger-scale computer game. Experiments demonstrate the long-term learning outcomes when agents with different OMIs oppose each other in a number of different games. For experimental purposes, Part III assumes that Algorithm 3.4 is the complete definition of the acting agent. However, it is possible for this decision-making loop to be embedded in traditional NPC control code to control certain classes of decisions, while other decisions are made by different decision-making processes. For NPCs defined by state machine-based scripting languages, for example, Algorithm 3.4 may control execution in one particular state, with other types of reasoning used in other states.

Compared to other learning algorithms such as neural networks and reinforcement learning, Algorithm 3.4 is a lightweight approach to learning about specific scenarios. In terms of memory requirements, storage of the values Ω and P $(B_t = B^C)$ is the main novel memory requirement. Most traditional agent-based approaches will already incorporate behaviour constructs in some form. Memory of the most recent executed behaviour B_{t-1} is also required, along with memory of the most recent payoff. The perception and learning updates (Eqs. 2.13 and 3.15) are constant time calculations, given that we assume a choice of only two behaviours.

3.5 Evolution

Evolutionary approaches such as genetic algorithms [8] and replicator equations [25] simulate the process of biological evolution by implementing concepts such as natural selection, reproduction and mutation.

Computational evolution has been incorporated in a number of recent commercial and experimental games. In *Creatures* [11], *Norns* can be cross-bred to evolve new kinds of *Norns*. Will Wright's *Spore* [1] is based on the concept of user-guided evolution, with players guiding a species through several stages of evolution.

Ken Stanley's research group at the University of Central Florida has produced a number of online games using evolutionary algorithms. *Galactic Arms Race* [13, 14] is a multiplayer online space combat game in which available weapons evolve in response to usage statistics. The *NERO Video Game* (short for Neuro-Evolving Robotic Operatives) [29, 30] permits user-guided evolution of combat robots.

In Part IV of this book we use replicator equations to explore the evolution of motives in societies of motivated agents. Replicator and replicator-mutator equations model selection and mutation. We use the theory of n-player social dilemma games to develop algorithms for the evolution of motivated agents. We first consider the structure of such games in Sect. 3.5.1 and the formalisms for evolution in Sect. 3.5.2. We then present algorithms for the evolution of motivated agents in Sect. 3.5.3.

3.5.1 Multiplayer Social Dilemma Games

Formally, in an n-player social dilemma game, each player again has two choices: B^C or B^D. Depending on the combination of choices made by all players, Player 1 is assigned a payoff value V^1, Player 2 a payoff value V^2 and so on up to Player n, assigned a payoff value V^n.

Suppose we now consider the n-player game from the perspective of a single player. The number of other players choosing B^C is denoted by n^C and the expected payoff to a player choosing B^C is $E(B^C)$. The payoff to players choosing B^D is $E(B^D)$. The total payoff to a player is defined in terms of the decisions made by the $n-1$ other players as follows:

$$E(B^C) = n^C R + (n-1-n^C)S, \qquad (3.16)$$

$$E(B^D) = n^C T + (n-1-n^C)P. \qquad (3.17)$$

The values of $E(B^C)$ and $E(B^D)$ at their end points are found by first setting $n^C = 0$ and then $n^C = n - 1$. The payoff to a single B^C chooser is $S(n - 1)$ and the payoff to the B^D choosers is $P(n-1)$. Likewise, the payoff to a single B^D chooser is $T(n-1)$ and the payoff to the B^C choosers is $R(n-1)$. We can see that the n-player game is simply a generalisation of the two-player game. In fact, we can think of one generation in an n-player game as a series of two-player games in which each of the n players takes on every other player and their payoff in each of these interactions is summed at the end of the generation. As a result, n-player games can be referred to as compound games.

3.5.2   Evolution in Multiplayer Social Dilemma Games

Now suppose that we do not care how many agents we have, but only about the fraction of each type of agent in the population. We can then construct a vector $\mathbf{x}_t$ stipulating the fractions of each type of agent in the society. Evolution can be modelled using a replicator equation to update $\mathbf{x}_t$ based on the fitness of each type of agent. Fitness $f^k(x)$ is dependent on the fraction of each type of agent in the population, but otherwise may be computed similarly to Eqs. 3.16 and 3.17.

The replicator equation itself is a deterministic, monotone, nonlinear and non-innovative game dynamic that we borrow from the field of evolutionary game theory. The most general continuous form is given by the differential equation:

$$\dot{x}^k = x^k[f^k(x) - \phi(x)], \tag{3.18}$$

where x^k denotes the fraction of the kth type of agent, $f^k(x)$ is the fitness of the kth type of agent and $\phi(x) = \sum_{k=1}^{K} x^k f^k(x)$ is the average population fitness, defined as the weighted average of the fitness of the K types in the population.

To simplify analysis, fitness is often assumed to depend linearly upon the population distribution, which allows the replicator equation to be written in the form:

$$\dot{x}^k = x^k[(\mathbf{Wx})^k - \mathbf{x}^T\mathbf{Wx}], \tag{3.19}$$

where $\mathbf{W}$ is a game, such as in Eq. 3.12, and holds the fitness information for the population in the form of payoff values.

3.5.3   Evolution of Motivated Agents

We now proceed to use the theory presented above in two algorithms to model the evolution of motivated agents. The first, Algorithm 3.5, assumes an objective definition of fitness, such that all agents are evaluated as fit in the same way, even if they have different motive profiles. An example of such a fitness function is a survival-related function. All agents are unfit if they die, regardless of their motive profile.

The second approach, in Algorithm 3.6, assumes a subjective definition of fitness, such that agents are evaluated as fit if they can satisfy their own motives, even if they do not maximise the objective fitness function of the scenario.

3.5.3.1 Evolution of Motivated Agents using Objective Fitness

Like motivated learning agents, we assume that agents in our evolutionary setting systematically misperceive a game. Their level of misperception is again a function of their OMI according to Eq. 2.13. In Algorithm 3.5, misperception influences behaviour, but does not influence fitness. Fitness is defined using an objective function $f^k(x)$(line 1). We use Eq. 3.18 as the basis for this approach. The fraction of each of K types of agents in the population is initialised in a vector $\mathbf{x}_0$ (line 2). Each agent type is assumed to have different OMI. The agents themselves are then initialised (line 3).

On each iteration (generation) of the algorithm the proportions of different types of agents are updated according to:

$$x_{t+1}^k = x_t^k + h\dot{x}_t^k \tag{3.20}$$

where h is a step size parameter. Smaller step sizes result in slower changes to the composition of the population, and a generally more stable population. Large step sizes can produce large fluctuations in the proportions of the different types of agents. The updated vector $\mathbf{x}_{t+1}$ must be normalised by zeroing any negative fractions and ensuring that $\sum_k x_{t+1}^k = 1$.

3.5.3.2 Evolution of Motivated Agents using Subjective Fitness

In this algorithm, misperception influences fitness and behaviour. Each agent has its own subjective fitness function. We use a variant of the replicator equation in Eq. 3.19 as the basis of this approach.

To construct an algorithm for the evolution of a society of motivated agents (Algorithm 3.6), we first select payoff constants T, R, S and P (line 1). These payoff constants represent the payoff that agents would receive in a two-player interaction. However, we use these to construct a compound game world $\mathbf{W}$ (line 2) for k different types of motivated agents. An agent type here is defined by its OMI. The total number of different types of agents (i.e., the maximum number of different OMIs among agents in the society) is K. The compound game world $\mathbf{W}$ in which motivated agents evolve is represented by a $2K$-by-$2K$ matrix concatenating the perceived two-player games of each of the K types of agents vertically, and then repeating this horizontally K times. These games are constructed using Eq. 2.13 to transform each incentive in the original game. The general form of $\mathbf{W}$ is given in Eq. 3.21. Specific examples will be given in Chap. 9.

$$\mathbf{W} = \begin{bmatrix} \hat{P}^1 & \hat{T}^1 & & \hat{P}^1 & \hat{T}^1 \\ \hat{S}^1 & \hat{R}^1 & \cdots & \hat{S}^1 & \hat{R}^1 \\ & & \cdot & & \\ \hat{P}^k & \hat{T}^k & & \hat{P}^k & \hat{T}^k \\ \hat{S}^k & \hat{R}^k & \cdots & \hat{S}^k & \hat{R}^k \\ & & \cdot & & \\ \hat{P}^K & \hat{T}^K & & \hat{P}^K & \hat{T}^K \\ \hat{S}^K & \hat{R}^K & \cdots & \hat{S}^k & \hat{R}^K \end{bmatrix}. \tag{3.21}$$

We then construct a vector $\mathbf{x}_t$ stipulating the fractions of each type of agent in the society (line 2). Differently from Algorithm 3.5, agents are further broken down by the proportion at any time choosing B^C and B^D as follows:

$$\mathbf{x}_t = \begin{bmatrix} F^1(B_t = B^C) \\ F^1(B_t = B^D) \\ \cdot \\ \cdot \\ F^k(B_t = B^C) \\ F^k(B_t = B^D) \\ \cdot \\ \cdot \\ F^K(B_t = B^C) \\ F^K(B_t = B^D) \end{bmatrix}. \tag{3.22}$$

$F^k(B_t = B^i)$ is the fraction of agents that are of type k (with OMI Ω^k) and will choose B^i. This means that $\sum_i F^k(B_t = B^i)$ is the fraction of agents of type k. The probability of an agent of a given type choosing B^C is

$$P^k(B_t = B^C) = \frac{F^k(B_t = B^C)}{\sum_i F^k(B_t = B^i)}, \tag{3.23}$$

and likewise for B^D.

A specific number of agents n is then chosen (line 3) and a society $\mathbf{A}$ of n agents created such that the distribution of agent OMIs conforms to the proportions initialised in $\mathbf{x}_0$. These agents may be motivated rule-based agents, learning agents, or any other kind of motivated agent in which motivation and behaviour are governed by an OMI.

The evolutionary process is governed by the replicator equation:

$$\dot{\mathbf{x}}_{t+1} = \mathbf{Q}\mathbf{x}_t[\mathbf{W}\mathbf{x}_t - \mathbf{x}_t^T\mathbf{W}\mathbf{x}_t]. \tag{3.24}$$

$\mathbf{Q}$ is a matrix of transition probabilities for the mutation of type k to type l. When $\mathbf{Q}$ is the identity matrix, Eq. 3.24 models evolution without mutation (without innovation), equivalently to Eq. 3.19. Once again, specific examples of $\mathbf{Q}$ are given in Chap. 9. After the agents execute their motivated behaviours (lines 5–6), the Euler method can again be applied to simulate the process of evolution (line 7). $\mathbf{x}_{t+1}$ must again be normalised by zeroing any negative fractions and this time ensuring that $\sum_k \sum_i F^k(B_t = B^i) = 1$ (line 8).

Algorithm 3.5 (*Evolving the proportions of agents with different motives when fitness is determined objectively*)

1.	Define an objective fitness function $f^k(x)$
2.	Initialise $\mathbf{x}_0$ for K types of agents
3.	Initialise n and a society $\mathbf{A}$ of n agents with correct fractions of each type.
4.	Repeat for each time t:
5.	Repeat for each A in $\mathbf{A}$:
6.	Execute a motivated behaviour
7.	Compute $\mathbf{x}_{t+1}$ using Equation 3.18 and 3.20.
8.	Normalise $\mathbf{x}_{t+1}$
9.	Create a new generation of agents in society $\mathbf{A}$ (and remove the old generation) to reflect new proportions in $\mathbf{x}_{t+1}$

Algorithm 3.6 (*Evolving the proportions of agents with different motives when fitness is determined subjectively*)

1.	Initialise T, R, P, S and matrix $\mathbf{W}$ using Equation 3.21
2.	Initialise $\mathbf{x}_0$ of the form in Equation 3.22 and $\mathbf{Q}$
3.	Initialise n and a society $\mathbf{A}$ of n agents with correct proportions of each type.
4.	Repeat for each time t:
5.	Repeat for each A in $\mathbf{A}$:
6.	Execute a motivated behaviour
7.	Compute $\mathbf{x}_{t+1}$ using update in Equation 3.24 and 3.20
8.	Normalise $\mathbf{x}_{t+1}$
9.	Create a new generation of agents in society $\mathbf{A}$ (and remove the old generation) to reflect new proportions in $\mathbf{x}_{t+1}$

Finally, the composition of $\mathbf{A}$ is updated by creating a new generation of agents with the appropriate new distribution of OMIs (line 9).

Evolutionary dynamics predict the equilibrium outcomes of a multi-agent system when the survival of agents in iterative game-play is determined by their fitness. Using this approach, the proportion of a given type of agent playing B^C or B^D will increase if its subjective incentive is greater than the average subjective incentive of the population. Conversely, the proportion of a given type of agent will decrease if its subjective incentive is less than the average subjective incentive of the population. Agent types with exactly average subjective incentive will neither increase nor decrease in number, unless a mutation occurs. Agents are thus considered to be fitter if they satisfy their own motives, regardless of whether or not they receive the highest explicit incentive as defined in the original game.

As with the algorithm for motivated learning agents, Algorithms 3.5 and 3.6 offer lightweight approaches to evolution compared to other evolutionary algorithms such as genetic algorithms. Only a pair of fractions $F^k(B_t = B^C)$ and $F^k(B_t = B^D)$ is stored for each type of motivated agent. The main characteristic of a genetic algorithm that is not present using replicator equations is the actual pairing of agents followed by crossover and mutation of their genetic structures. However, the proposed approach avoids the associated cost to store these genetic structures and the computational complexity of the crossover processes.

3.6   Summary

In summary, this chapter has presented algorithms for motivated rule-based agents and motivated learning agents, as well as algorithms and for controlling crowds or the evolution of societies of motivated agents. The algorithms are designed to be lightweight in terms of their memory and computational requirements, and are flexible enough to be incorporated in rule-based or state machine architectures that traditionally have been used in game development.

Part II of this book will describe how the models of motivation introduced in Chap. 2 can be incorporated in the motivated rule-based agents introduced in Sect. 3.2.2. The resulting agents are then demonstrated by reproducing three canonical human experiments with artificial agents. Part III examines how models of motivation can be incorporated in the motivated rule-based, crowds of agents and learning agents. Finally, Part IV will examine the evolution of societies of motivated agents using the motivated evolutionary algorithm introduced in Sect. 3.5.3.

References

1. Spore (Maxis 2008), www.spore.com
2. D. Acemoglu, M. Yildiz, Evolution of perceptions and play. Massachusetts Institute of Technology, Department of Economics, Working Paper 01-36 (2001)
3. T. Borgers, R. Sarin, Learning through reinforcement and replicator dynamics. J. Econ. Theory **77**, 1–14 (1997)
4. A. Colman, *Game Theory and Experimental Games: The Study of Strategic Interaction* (Pergamon Press, Oxford, England, 1982)
5. E. Dekel, J. Ely, O. Ylankaya, Evolution of preferences. Rev. Econ. Stud. **74**, 685–704 (2007)
6. P. Fishburn, Lexicographic orders, utilities and decision rules: a survey. Manag. Sci. **20**, 1442–1471 (1974)
7. G. Gigerenzer, P. Todd, *Simple Heuristics That Make us Smart* (Oxford University Press, NY, 1999)
8. D. Goldberg, *Genetic Algorithms in Search, Optimisation and Machine Learning* (Addison-Wesley Professional, Reading, MA, 1989)

9. T. Graepel, R. Herbrich, J. Gold, Learning to fight, in *Proceedings of the International Conference on Computer Games: Artificial Intelligence, Design and Education* (2004)
10. J. Graham, J. Starzyk, D. Jachyra, Opportunistic behavior in motivated learning agents. IEEE Trans. Neural Netw. Learn. Syst. **26**, 1735–1746 (2014)
11. S. Grand, *Creation: Life and How To Make It* (Harvard University Press, 2003)
12. O. Guillermo, *Game Theory* (Academic Press, San Diego, CA, 1995)
13. E. Hastings, R. Guha, K. Stanley, Evolving content in the Galactic Arms Race video game, in *Proceedings of the IEEE Symposium on Computational Intelligence in Games* (Milano, 2009), pp. 241–248
14. E. Hastings, R. Guha, K. Stanley, Automatic content generation in the Galactic Arms Race video game. IEEE Trans. Comput. Intell. AI Games **1**, 245–263 (2009)
15. J. Heckhausen, H. Heckhausen, *Motivation and Action* (Cambridge University Press, New York, NY, 2010)
16. D. Kahneman, A. Tversky, Prospect theory: an analysis of decision under risk. Econometrica **47**, 263–292 (1979)
17. R.L. Keeney, H. Raiffa, *Decisions with Multiple Objectives: Preferences and Value Tradeoffs* (Wiley, New York, 1976)
18. D. Kehoe, Designing artificial intelligence for games (part 1) (2012), https://software.intel.com/en-us/articles/designing-artificial-intelligence-for-games-part-1. Accessed 27 Aug 2014
19. M. Mateas, Expressive AI: games and artificial intelligence, in *Proceedings of the Level Up: Digital Games Research Conference* (Utrecht, Netherlands, 2003)
20. S. Meier, Civilization (2K Games, 1991), http://www.civilization.com/en/home
21. K. Merrick, M.L. Maher, *Motivated Reinforcement Learning: Curious Characters for Multiuser Games* (Springer-Verlag, Berlin, 2009)
22. K. Merrick, The role of implicit motives in strategic decision-making: computational models of motivated learning and the evolution of motivated agents. Games **6**, 604–636 (2015). (Special Issue on Psychological Aspects of Strategic Choice)
23. Microsoft, Drivatar in Forza Motorsport (2014), http://research.microsoft.com/en-us/projects/drivatar/forza.aspx. Accessed 31 Oct 2014
24. N.J. Nilsson, Introduction to machine learning (1996), http://ai.stanford.edu/people/nilsson/mlbook.html. Accessed Jan 2006
25. M. Nowak, *Evolutionary dynamics: exploring the equations of life* (Belknap Press, Canada, 2006)
26. C.W. Reynolds, Flocks, herds and schools: a distributed behavioral model. Comput. Gr. **21**, 25–34 (1987). (SIGGRAPH 87 Conference Proceedings)
27. R. Saunders, J.S. Gero, Curious agents and situated design evaluations. Artif. Intell. Eng. Des. Anal. Manuf. **18**, 153–161 (2004)
28. G. Seemann, D. Bourg, *AI for Game Developers* (O'Reilly Media Inc, 2004)
29. K. Stanley, R. Cornelius, R. Miikkulainen, T. D'Silva, A. Gold, Real-time learning in the NERO video game, in *Proceedings of the Annual AAAI Conference on Artificial Intelligence and Interactive Digital Entertainment* (2005)
30. K. Stanley, I. Karpov, R. Miikkulainen, A. Gold, The NERO video game, in *Proceedings of the Annual AAAI Conference on Artificial Intelligence and Interactive Digital Entertainment* (2006), pp. 151–152
31. J. Von Neumann, O. Morgenstern, *Theory of Games and Economic Behavior* (Princeton University Press, Princeton, NJ, 1953)
32. M. Wang, K. Hipel, N. Fraser, Modeling misperceptions in games. Behav. Sci. **33**, 207–223 (1988)
33. G. Yannakakis, Game AI revisited, in *Proceedings of the Ninth Conference on Computing Frontiers*, (Cagliari, Italy, 2012), pp. 285–292
34. N. Yee, Motivations of play in online games. Cyberpsychol. Behav. **9**, 772–775 (2007)

Part II
Comparing Human and Artificial Motives

Chapter 4
Achievement Motivation

Part II of this book describes how the models of motivation introduced in Part I can be used to create a number of specific motivation subtypes that have previously been observed in humans. These are demonstrated by reproducing three canonical human experiments with artificial agents. In this chapter, the focus is on achievement motivation. Four subtypes of achievement motivation are modelled. These models are embedded in agents playing the ring-toss game. We demonstrate the similarities that can be observed between motivated game-playing agents and humans playing the same game. We also compare our achievement-motivated agents to agents using other models of achievement motivation.

4.1 Scenarios and Mini-games for Motivated Agents

The focus of this chapter, and subsequent chapters, is on motivated agents playing individual, self-contained games. This permits us to conduct an empirical analysis of the behaviour of different types of agents under controlled and constrained conditions. Mini-games or subgames are often embedded in larger virtual worlds and multiplayer online games. Isolation of mini-games within a larger game world, for example is common to give game designers greater control over the behaviour of non-player characters as each character is confined to a certain area. Mini-games can be isolated by terrain conditions or levelling requirements or because non-player characters lack the sensory mechanisms to reason about events beyond their immediate location. This chapter and the next focus on three specific games: ring-toss, roulette and the prisoners' dilemma.

In this chapter we examine the goal-selecting behaviour of agents with a single type of motivation: achievement motivation. We examine these agents playing a ring-toss game. This game was originally chosen by psychologists as appropriate for investigating achievement motivation in humans [1], because a player can stand at different distances from the spike. Tossing a ring from a given distance with the

© Springer International Publishing AG 2016
K.E. Merrick, *Computational Models of Motivation for Game-Playing Agents*,
DOI 10.1007/978-3-319-33459-2_4

aim of getting it over the spike constitutes a goal. The game thus comprises a series of goals of different difficulty according to the distance from the spike that the player chooses to stand. Psychologists hypothesize that individuals with different levels of achievement motivation will choose different distances from which to toss their ring. The ring-toss game is introduced in further detail in Sect. 4.2.

In Sect. 4.3 we model four subtypes of achievement-motivated agents corresponding to subtypes identified in humans, and study these agents playing the ring-toss game. The aims of the study are twofold. In Sect. 4.4.1 we compare the goal-selecting characteristics of achievement-motivated rule-based agents to humans playing the ring-toss game. In conducting this experiment, we are interested in determining whether it is possible to create artificial agents with similar goal-selecting characteristics as humans, at least within the controlled circumstances of a particular game.

In Sects. 4.4.2 and 4.4.3 we compare the goal-selecting characteristics of our achievement-motivated agents to those of two other models of achievement motivation: Atkinson's risk-taking model and an alternative computational model proposed by Simkins et al. [5]. We conclude that our model provides more flexibility to achieve different behavioural variants observed in humans.

We conclude in Sect. 4.5 with a discussion of the implications of achievement-motivated rule-based agents for non-player characters in general.

4.2 The Ring-Toss Game

The ring-toss game (also called quoits) involves throwing a metal, rope or rubber ring over a set distance to land over a spike, as shown in Fig. 4.1. The task of tossing the ring over the spike is classified as an achievement task, which succeeds when the ring lands over the spike. The difficulty of the task—and thus the probability of success—is related to the distance the player stands from the spike.

In psychology, a ring-toss experiment was originally designed to verify theories of achievement motivation in humans [1]. The ring-toss experiment has two phases. The first phase involves determining the person's level of achievement motivation using specially designed tests. The second phase is the ring-toss phase. In humans, level of achievement motivation can be measured in terms of tendency to approach success or avoid failure. Tendency to approach success can be measured using the projective measure of need achievement or the thematic apperception test (TAT), while tendency to avoid failure can be measured using the Mandler-Sarson test of test anxiety [2]. Once a person has taken these tests, their results can be used to compute their level of achievement motivation. This level can then be linked experimentally to characteristic behavioural response types during the ring-toss phase of the experiment.

Atkinson and Litwin [1] conducted an investigation of the effects of achievement motivation in a ring-toss experiment. Individuals' tendency to approach success or

Fig. 4.1 The ring-toss game. Image from [4]

avoid failure was gauged using the projective test of need achievement and Mandler-Sarason tests. Individuals were then broken into four groups as follows:

- **H-L:** high motivation to approach success and low motivation to avoid failure,

- **H-H:** high motivation both to approach success and to avoid failure,

- **L-L:** low motivation both to approach success and to avoid failure,

- **L-H:** low motivation to approach success and high motivation to avoid failure.

Atkinson and Litwin [1] had 45 participants in their experiment and each was allowed ten opportunities to toss a ring at a peg from a distance of their choice in the range of zero to fifteen feet (approximately 4.57 m). Of the subjects, 13 were classified as H-L, 10 as H-H, nine as L-L and 13 as L-H. Atkinson and Litwin [1] collated their results for each motivation type in a series of three range brackets roughly corresponding to 'easy', 'moderate' and 'difficult' goals. They in fact proposed a number of different definitions for easy, moderate and difficult ring-toss goals, which result in different range brackets. We chose the third of their definitions, which takes into account both distance to target and the distribution of those distances about the median. Their results collated using this definition are shown in Table 4.1.

Table 4.1 Percentage of human participants choosing 'easy', 'moderate' and 'difficult' throwing distances, grouped by achievement motivation subtype

Range bracket (m)	H-L (%)	H-H (%)	L-L (%)	L-H (%)
0.00–2.00 (easy)	11.0	26.0	18.0	32.0
2.25–3.50 (moderate)	82.0	60.0	58.0	48.0
3.75–4.50 (difficult)	7.0	14.0	24.0	20.0

Numbers from [1]

In the next section we describe how agents can be constructed to reproduce the behavioural characteristics of individuals with each of these achievement motivation subtypes.

4.3 Modelling Achievement-Motivated Rule-Based Agents

The experiments in this chapter use the motivated rule-based agent in Algorithm 3.1 to create H-L, H-H, L-L and L-H achievement-motivated rule-based agents (called H-L AchAgents; H-H AchAgents, L-L AchAgents and L-H AchAgents respetively). For the purpose of simulating multiple agents, sensors and behaviours are not implemented. Only the components of the algorithm necessary for collecting data are implemented. We implemented these experiments in MATLAB. The specific settings used are as follows:

- Line 1: Motivation is modelled using Eq. 2.6 with parameters in Table 2.2;

- Line 3: In this study we do not model sensors explicitly;

- Line 4: Nineteen condition-goal-behaviour tuples were used. C = true for all tuples as we do not model sensors. Goals range from throws at 0–4.5 m in increments of 0.25 m. We make an arbitrary task-based estimate of the probability of success for tosses from different distances. $P^s(G^1)$, is assigned the value 0.99. $P^s(G^{19})$ is assigned the value 0.10. Other probabilities are distributed uniformly within this range. These probabilities are the same for all agents. Table 4.2 labels and summarizes these goals. The agents are assumed to have behaviours B^1–B^{19} that comprise sequences of actions for moving to the correct position and throwing the ring at the spike, although these are not explicitly programmed in this experiment.

- Line 5: Goal selection is modelled using Eq. 2.14.

To create the four agent subtypes corresponding to the H-L, H-H, L-L and L-H individuals identified by Atkinson and Litwin [1], we need to determine values for the parameters M_{ach}^+, M_{ach}^-, S_{ach}, ρ_{ach}^+ and ρ_{ach}^-. To permit variation and diversity, we propose that this be done by identifying ranges for each parameter and then randomly select values from those ranges. According to the definitions of H-L, H-H,

Table 4.2 Ring-toss goals and their probability of success in the computational experiments

Goal	Distance from target (m)	$P^s(G)$
G^1	0.00	0.990
G^2	0.25	0.941
G^3	0.50	0.891
G^4	0.75	0.842
G^5	1.00	0.792
G^6	1.25	0.743
G^7	1.50	0.693
G^8	1.75	0.644
G^9	2.00	0.594
G^{10}	2.25	0.545
G^{11}	2.50	0.496
G^{12}	2.75	0.446
G^{13}	3.00	0.397
G^{14}	3.25	0.347
G^{15}	3.50	0.298
G^{16}	3.75	0.248
G^{17}	4.00	0.199
G^{18}	4.25	0.149
G^{19}	4.50	0.100

L-L and L-H individuals, M_{ach}^+ and M_{ach}^- are key to creating agents with different subtypes of achievement motivation. Accordingly, we do this as follows:

- **H-L AchAgents** have a 'low' turning point M_{ach}^+ for approach motivation and a 'high' turning point M_{ach}^- for avoidance motivation,

- **H-H AchAgents** have a 'low' turning points for both approach and avoidance,

- **L-L AchAgents** have a 'high' turning points for both approach and avoidance,

- **L-H AchAgents** have a 'high' turning point for approach motivation and a 'low' turning point for avoidance motivation.

Given that probability of success must fall in the range [0, 1], we interpret the qualitative terms 'low' and 'high' used above as referring to positions in this range. We propose that a low range value for M_{ach}^+ falls in the interval [0, 0.1], while a high range value for M_{ach}^+ falls in the interval [0.1, 0.2]. A low range value for M_{ach}^- falls in the interval [0.8, 0.9], and a high range value falls in the interval [0.9, 1.0]. We do not consider cases where $M_{ach}^+ > M_{ach}^-$, as Atkinson and Litwin [1] discovered that failure motivated individuals to do not exhibit preferences for very easy or very difficult tasks to the extent predicted by this model. A summary of the suggested values for M_{ach}^+ and M_{ach}^- is given in the second and third rows of Table 4.3.

Table 4.3 Achievement motivation parameters and suggested value ranges

Parameter	H-L AchAgents	H-H AchAgents	L-L AchAgents	L-H AchAgents
S_{ach}	1.0	1.0	1.0	1.0
M_{ach}^{+}	[0.1, 0.2]	[0.1, 0.2]	[0, 0.1]	[0, 0.1]
M_{ach}^{-}	[0.8, 0.9]	[0.9, 1.0]	[0.8, 0.9]	[0.9, 1.0]
ρ_{ach}^{+}	[22.3, 76.0]	[11.3, 94.7]	[0.0, 93.6]	[2.7, 91.3]
ρ_{ach}^{-}	[2.0, 66.0]	[0.0, 67.7]	[0.0, 67.0]	[0.0, 84.0]

S_{ach} was fixed at an arbitrarily chosen value. M_{ach}^{+} and M_{ach}^{-} were chosen based on the psychological literature. ρ_{ach}^{+} and ρ_{ach}^{-} were chosen experimentally

The value for $S_{ach} = 1.0$ was used for all agent subtypes. We fix this for two reasons. First, it reduces the complexity of the search for appropriate parameter values because the values of ρ_{ach}^{+} and ρ_{ach}^{-} also influence the maximum strength of the motivational tendency. Secondly, it permits us to reserve the S parameter for controlling the relative strength of motives when there is more than one motive. We do this in Chap. 5.

This leaves us to determine ranges for ρ_{ach}^{+} and ρ_{ach}^{-}. As the psychology literature does not suggest how these should be chosen for artificial agents, we conduct an experiment search for the lower and upper bounds of these ranges for each agent subtype. First, we select fifty-five potential combinations of upper and lower bounds from the interval [0, 100]. These combinations are enumerated in Table 4.4. Gradients greater than 100 do not significantly change the shape of the motivation curve (which becomes increasingly close to vertical with higher values), so we limit the space of consideration in this way. We consider each of the fifty-five combinations of upper and lower bounds for both ρ_{ach}^{+} and ρ_{ach}^{-}, giving a total of 3,025 combinations of upper and lower bounds to explore.

Using each combination, 1,000 agents of each subtype were created by randomising parameters from a uniform distribution over the selected range. These agents were then allowed to select one goal distance from which to toss a ring at the

Table 4.4 Values considered for the upper and lower bounds of ρ_{ach}^{+} and ρ_{ach}^{-}

	Upper bound									
	10	20	30	40	50	60	70	80	90	100
Lower bound	0	10	20	30	40	50	60	70	80	90
		0	10	20	30	40	50	60	70	80
			0	10	20	30	40	50	60	70
				0	10	20	30	40	50	60
					0	10	20	30	40	50
						0	10	20	30	40
							0	10	20	30
								0	10	20
									0	10
										0

spike. That is, each agent made one pass through the intelligence loop in Algorithm 3.1. The number of agents choosing easy, moderate and difficult goals was recorded. The upper and lower bounds that resulted in the greatest similarity between proportions of agents and humans choosing easy, moderate and difficult goals were then identified and recorded. Table 4.1 was used as the reference point for human data. This constituted one epoch. The experiment was repeated thirty times and the best performing lower and upper bounds from each epoch recorded. These bounds were averaged to produce the suggested parameter value ranges shown in the last two lines of Table 4.3.

We see from Table 4.3 that the best performing ranges of ρ_{ach}^{+} have, on average, higher upper bounds than ρ_{ach}^{-}, and generally higher lower bounds as well. The implication is that agents with the highest values of ρ_{ach}^{+} and lowest values of ρ_{ach}^{-} will have $\rho_{ach}^{+} > \rho_{ach}^{-}$. That is, there will be clear cases where the gradient of approach is greater than the gradient of avoidance. There will also be cases where the converse is true. In the next section, AchAgents using the value ranges in Table 4.3 are compared to humans. We also discuss agents using Atkinson's original risk-taking model and an existing alternative computational model of achievement motivation.

4.4   AchAgents Playing the Ring-Toss Game

In this experiment, a thousand of each subtype of achievement-motivated agent was created using the parameter values in Table 4.3. Each simulated agent was allowed to select one goal distance from which to toss a ring at the spike. That is, each agent made one pass through the intelligence loop in Algorithm 3.1. The distance selected was recorded for each agent and the results collated. These results for were then compared to experimental results with humans reported by Atkinson and Litwin [1].

4.4.1   Comparison to Humans

Figure 4.2 collates the experimental results for the 4,000 AchAgents and Table 4.5 summarizes the salient statistics for these agents. Table 4.5 shows the average ring-toss distance selected by H-L AchAgents is 2.79 m. This corresponds most closely to G^{12} ($P^{s} = 0.446$). H-L AchAgents show the most focused goal selection, as indicated by the lowest recorded standard deviation of 0.514 in Table 4.5. This is because the average distance between the turning points M_{ach}^{+} and M_{ach}^{-} is the smallest for these agents (0.7 m on average). This has the effect of narrowing the range of goal probabilities that are likely to be selected by these agents. This tendency to focus on a narrow range of goals of intermediate-high difficulty is in keeping with the results reported by Atkinson and Litwin [1] (pp. 55–56) for

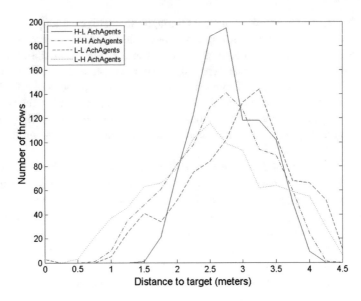

Fig. 4.2 Raw results when 4,000 AchAgents participate in the ring-toss experiment. 1,000 each of H-L, H-H, L-L and L-H AchAgents were used. Each agent had one throw

Table 4.5 Summary of salient statistics for Fig. 4.2

	H-L AchAgents	H-H AchAgents	L-L AchAgents	L-H AchAgents
Average distance selected (m)	2.789	2.651	2.945	2.567
Standard deviation	0.514	0.707	0.796	0.900
95 % confidence interval	±0.032	±0.044	±0.049	±0.056

humans. They interpret this as meaning that individuals high in achievement motivation desire to focus on goals that they believe will be challenging.

H-H AchAgents select an average ring-toss distance of 2.65 m (also closest to G^{12}) while L-L AchAgents select an average ring-toss distance of 2.95 m (closest to G^{13}). Both these types of agents exhibit less focused goal selection than H-L AchAgents, indicated by higher standard deviations of 0.707 and 0.796 respectively. This higher standard deviation is in keeping with the results for humans. H-H AchAgents select goals with smaller distances to the target on average than L-L AchAgents. The difference, which is statistically significant at the 95 % confidence level, is caused by the lower average values of M_{ach}^{+} and M_{ach}^{-} for L-L AchAgents. It is not clear from existing descriptions of human experiments whether this result can be observed in humans.

Table 4.6 Comparison of AchAgents and humans in twelve categories

Range bracket (m) and corresponding goals	H-L	H-H	L-L	L-H
	Humans AchAgents Z-Value	Humans AchAgents Z-Value	Humans AchAgents Z-Value	Humans AchAgents Z-Value
0.00–2.00 $(G^1$–$G^9)$	11.0 % 9.7 % −0.468	26.0 % 23.7 % −0.514	18.0 % 16.0 % −0.494	32.0 % 31.4 % −0.139
2.25–3.50 $(G^{10}$–$G^{15})$	82.0 % 84.4 % 0.582	60.0 % 67.8 % 1.583	58.0 % 64.3 % 1.191	48.0 % 53.7 % 1.225
3.75–4.50 $(G^{16}$–$G^{19})$	7.0 % 5.9 % −0.496	14.0 % 8.5 % −1.832	24.0 % 19.7 % −0.976	20.0 % 14.9 % −1.512

The percentage of humans and agents selecting goals in each category is shown. The critical z-value for this test is ±1.96 at the 95 % confidence level

L-H AchAgents select an average ring-toss distance of 2.567 m. This corresponds most closely to G^{11}. Unlike H-L AchAgents, L-H AchAgents have much less focused goal selection. This is indicated by a higher standard deviation of 0.900. The difference in focus is also in keeping with the results for humans reported by Atkinson and Litwin [1]. The high standard deviation for throw distances in L-H AchAgents is caused by the larger average difference between M_{ach}^{+} and M_{ach}^{-} (0.9 on average).

When the three 'difficulty' brackets identified by Atkinson and Litwin [1] (see Table 4.1) are multiplied by the four achievement motivation types (H-L, H-H, L-L and L-H), this gives a total of twelve experimental categories as shown in Table 4.6.

Table 4.6 collates the results for AchAgents into these twelve categories and shows the z-value for the two results at the 95 % confidence interval. The critical z-value for a two-tailed z-test of two proportions at the 95 % confidence level is ± 1.96. Table 4.6 shows that AchAgents produce statistically similar results to those of human studies in all twelve experimental categories.

4.4.2 Comparison to Atkinson's Risk-Taking Model

Suppose that instead of using Eq. 2.6 to compute the maximally motivating goal in line 5 of Algorithm 3.1, we use Atkinson's risk-taking model (RTM) as defined in Eq. 1.5. Then in line 5 of Algorithm 3.1 we use:

$$\left(T_{ach}^{res} \circ P^s\right)(G) = \left(M^s - M^f\right)\left(P^s(G) - [P^s(G)]^2\right) \tag{4.1}$$

Using this approach:

- **H-L RTMAgents** have a 'high' value for approach motivation M^s and a 'low' value for avoidance motivation M^f,

- **H-H RTMAgents** have 'high' values for both approach and avoidance motivation,

- **L-L RTMAgents** have 'low' values for both approach and avoidance motivation,

- **L-H RTMAgents** have a 'low' value for approach motivation and a 'high' value for avoidance motivation.

Again, a thousand of each type of these agents were initialised with parameter values within specific ranges to achieve each characteristic motivation type. The parameter ranges used are shown in Table 4.7.

Figure 4.3 collates the experimental results for the 4,000 RTMAgents and Table 4.8 summarizes the salient statistics for these agents. As discussed in Chap. 2,

Table 4.7 Achievement motivation parameters and their experimental value ranges for agents using Atkinson's risk-taking model (RTM)

Parameter	H-L RTMAgents	H-H RTMAgents	L-L RTMAgents	L-H RTMAgents
M^s	[0.9, 1.0]	[0.9, 1.0]	[0.8, 0.9]	[0.8, 0.9]
M^f	[0.0, 0.1]	[0.2, 0.3]	[0.0, 0.1]	[0.2, 0.3]

The parameter values for each agent are randomly generated within the given range

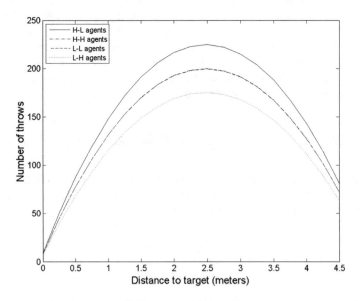

Fig. 4.3 Raw results when 4,000 RTMAgents participate in the ring-toss experiment. 1,000 each of H-L, H-H, L-L and L-H RTMAgents were used. Each agent had one throw

Table 4.8 Comparison of RTMAgents and humans

Range bracket (m)	H-L	H-H	L-L	L-H
	Humans RTMAgents Z-Value	Humans RTMAgents Z-Value	Humans RTMAgents Z-Value	Humans RTMAgents Z-Value
0.00–2.00 $(G^1$–$G^9)$	11.0 % 0.0 % −3.890*	26.0 % 0.0 % −5.467*	18.0 % 0.0 % −4.219*	32.0 % 0.0 % −7.037*
2.25–3.50 $(G^{10}$–$G^{15})$	82.0 % 100.0 % 5.071*	60.0 % 100.0 % 7.071*	58.0 % 100.0 % 6.917*	48.0 % 100.0 % 9.58*
3.75–4.50 $(G^{16}$–$G^{19})$	7.0 % 0.0 % −4.591*	14.0 % 0.0 % −3.880*	24.0 % 0.0 % −4.954*	20.0 % 0.0 % −5.375*

The percentage of humans and agents selecting goals in each category is shown. The critical z-value for this test is ±1.96 at the 95 % confidence level
*Indicates a statistically significant difference in results

Atkinson's RTM enforces a maximum motivational tendency at $P^s(G) = 0.5$ regardless of the values chosen for other parameters. This corresponds to G^{10} and G^{11} (2.25 and 2.50 m) in our simulation. The result is that all agents choose these moderate difficulty goals, as shown in Table 4.8, regardless of the values of other parameters. Because $[P^s(G)]^2$ and $P^s(G)$ have the same coefficients $(M^s - M^f)$, we cannot easily control the position of the maximum without modifying the model in some way through the addition of more constants. While this is possible, it is easier, and more intuitive, to consider an alternative to the quadratic model to gain the fine control that we desire.

It should be noted that this experiment does not negate Atkinson's model, which has been enormously successful in describing human motivation. Rather, it demonstrates that this model is inappropriate for use in artificial agents and supports the development of alternative models for this purpose.

4.4.3 Comparison to a Model of Achievement Motivation with Learning

One such alternative model of achievement motivation was proposed by Simkins et al. [3, 5]. These agents are denoted by SimAgents in Table 4.9. SimAgents use a combination of computational achievement motivation and reinforcement learning (RL) to model agent personality. The difference between SimAgents and the AchAgents described in this chapter is that SimAgents have a RL phase before the ring-toss experiment is conducted. In the RL phase, SimAgents play the ring-toss game and are rewarded when they succeed in a throw from a distance that satisfies their individual achievement profile. Over time each agent converges on a distance

Table 4.9 Adapted from Simkins et al. [5]: Comparison of SimAgents and humans

Range bracket (m)	H-L Humans SimAgents Z-Value	H-H Humans SimAgents Z-Value	L-L Humans SimAgents Z-Value	L-H Humans SimAgents Z-Value
0.00–2.00 (G^1–G^9)	11.0 % 7.7 % –0.914	26.0 % 14.0 % –2.121*	18.0 % 5.6 % –2.578*	32.0 % 8.5 % –4.714*
2.25–3.50 (G^{10}–G^{15})	82.0 % 75.4 % –1.300	60.0 % 69.0 % 1.330	58.0 % 74.4 % 2.326*	48.0 % 80.0 % 5.375*
3.75–4.50 (G^{16}–G^{19})	7.0 % 16.9 % 0.418	14.0 % 17.0 % 0.586	24.0 % 20.0 % –0.648	20.0 % 11.5 % –1.881

The percentage of humans and agents selecting goals in each category is shown. The critical z-value for this test is ±1.96 at the 95 % confidence level
*Indicates a statistically significant difference in results

that accumulates the most reward, and thus best satisfies its achievement profile. Only at this point are the results recorded. Simkins et al. used forty-nine different agents with different achievement profiles.

Table 4.9 shows the results reported by Simkins et al. [5] compared to Atkinson and Litwin's [1] human results. Simkins et al. [5] used confidence intervals to compute the statistical difference between their model and human performance. Using that approach, SimAgents have statistically different performance to humans in eight of the twelve experimental categories. For consistency, Table 4.9 shows the z-values for the Simkins et al. [5] model. However, the results still show a statistical difference in five of the twelve categories.

4.5   Conclusion

This chapter has introduced four subtypes of the achievement-motivated agents: H-L AchAgents, H-H AchAgents, L-H AchAgents and L-L AchAgents. This was done by tuning the turning point and slope parameters of the sigmoid function used to model motivation. The resulting agents were then used to replicate a human experiment in which participants were asked to select positions from which to throw rings in a ring-toss game.

The main assumption of our experimental design was that agents can associate a probability of success with each of nineteen discrete distances from the target. In this work, these probabilities were chosen so that the probability of success for zero distance was an arbitrary high number (0.99) and decreased in a linear way down to an arbitrary low number (0.1) at the maximum allowable distance to target. It was further assumed that all agents associate the same probability of success with a given distance to target. This represents a form of mastery-oriented achievement motivation. In practice we don't know how the human participants in the original experiment evaluated the probability of success of a throw. If this differs from our assumptions then it is likely to impact the parameter ranges required to achieve a match between agent and human behaviour.

The assumptions above notwithstanding, we draw two main conclusions from our experiments. First, it is possible to develop agents with goal-selection characteristics that are statistically similar to trends observed in human behaviour under certain constrained conditions of the ring-toss game. In general, achievement-motivated agents select goals of intermediate to high difficulty. This result reflects two of the characteristics of achievement-motivated individuals identified in Chap. 1 (Table 1.2). That is, they prefer choosing and accomplishing challenging goals, and are willing to take calculated risks. In this case, the goals are challenging because the agents stand back from the spike, but the risk is mitigated by not standing too far from the spike. More specifically, we are able to develop subtypes of achievement-motivated agents that reflect some of the subtle variations observed in achievement-motivated behaviour in humans.

Our second conclusion is that the sigmoid-based model in Chap. 2 provides a more accurate approach to modelling achievement motivation for artificial agents than a direct adaptation of Atkinson's risk-taking model. It also improves on the learning-based computational model of achievement motivation proposed by Simkins et al. [5].

When we reflect on these conclusions in the context of non-player characters, we see that there is evidence to support the use of computational motivation in designing non-player characters with more diverse and human-like behavioural characteristics, at least in certain constrained scenarios. Using the agent model in this chapter applied to agents selecting distances from which to shoot or throw at a target; for example, the agents will distribute themselves at different distances from the target. The same agents can be applied to other decision-making scenarios and will have the same characteristics as a result of their motives. For example, if given a choice of enemies with different abilities, some will choose to fight moderately skilled opponents. The others will attack the most skilled opposition or mop up the stragglers. If given a choice of building tasks requiring raw materials of different rarity, some will focus on tasks with moderately hard to find materials, while the others will allocate themselves to the easier or more difficult tasks.

In the experiments in this chapter each agent was evaluated after one pass through its intelligence loop. Of course, even on multiple passes through the loop, the decision made by a single agent would be the same unless the perceived probability of success at a goal changed in some way. Diversity in a society of motivated rule-based agents thus comes from having many agents with different values of their motivation parameters. In Part III we will consider agents that learn and change their decision-making characteristics over time.

References

1. J.W. Atkinson, G.H. Litwin, Achievement motive and test anxiety conceived as motive to approach success and motive to avoid failure. J. Abnorm. Soc. Psychol. **60**, 52–63 (1960)
2. G. Mandler, S.B. Sarason, A study of anxiety and learning. J. Abnorm. Soc. Psychol. **47**, 166–173 (1952)
3. K. Merrick, A computational model of achievement motivation for artificial agents, in *Proceedings of the Autonomous Agents and Multi-Agent Systems (AAMAS 2011)*, Taiwan, 2011, pp. 1067–1068
4. K. Merrick, K. Shafi, Achievement, affiliation and power: motive profiles for artificial agents. Adapt Behav **19**, 40–62 (2011)
5. C. Simkins, C. Isbell, and N. Marquez, Deriving behavior from personality: A reinforcement learning approach, in *Proceedings of the International Conference on Cognitive Modelling*, Philadelphia, PA, 2010, pp. 229–234

Chapter 5
Profiles of Achievement, Affiliation and Power Motivation

In this chapter, three profiles of achievement, affiliation and power motivation are introduced for two different games: roulette and the prisoners' dilemma game. These profiles are tuned for agents playing roulette using a genetic algorithm, and then further examined in agents playing the prisoners' dilemma game. We again demonstrate the similarities that can be observed between motivated game-playing agents and humans playing the same games. We show that we can create a diverse group of agents that compete or cooperate in different ways when playing games, and that models tuned on one game can be used in agents playing another game.

5.1 Evolving Motivated Agents to Play Multiple Games

This chapter continues our examination of the behaviour of artificial agents using computational models of motivation when compared to humans in game-playing scenarios. In this chapter we examine the goal-selecting behaviour of agents with a profile of three motivations: achievement, affiliation and power motivation. The focus of this chapter is again on motivated agents playing an individual, self-contained game. In Sect. 5.2 the game studied is roulette. In Sect. 5.3 the game studied is the canonical prisoners' dilemma from game theory [6].

Roulette was originally chosen by psychologists to verify theories of achievement, affiliation and power motivation in humans [4]. Because a player can place different types of bets with different expected returns, the game defines a series of goals of different incentive. This property of roulette also satisfies a key requirement for the application of motivated rule-based agents, making it possible to reproduce the experiment for artificial agents. In addition, because players can see each other's bets, there is also an element of competition involved. Psychologists hypothesize that individuals who are stronger or weaker in different types of motivation will choose to place different types of bets, either to mitigate their own risks or to compete with the bets placed by others.

K.E. Merrick, *Computational Models of Motivation for Game-Playing Agents*, DOI 10.1007/978-3-319-33459-2_5

The prisoners' dilemma game, while having its own story, is a useful study because it can be represented mathematically in an abstract manner. This permits other scenarios to be identified that have the same properties. We can thus extend the results from this chapter beyond a single game, and into any scenario that conforms to the assumptions of this scenario. We do this in Chap. 6.

The primary aim of the studies in this chapter is to compare the goal-selecting characteristics of game-playing agents using a motive profile to humans playing the same game. In addition, we examine how well artificial agents with motive profiles developed for one game can transfer to play a second game. Specifically, Sect. 5.2 develops motivated agents to play roulette, which are then examined in Sect. 5.3 playing the prisoners' dilemma game. Section 5.2 includes a description of the use of a genetic algorithm to tune motivation parameters to capture behavioural characteristics identified in human experiments.

We conclude in Sect. 5.4 with a discussion of the implications of motivated rule-based agents for non-player characters in general.

5.2   Roulette

European roulette is a casino game in which players bet on the outcome of a ball spinning around a numbered wheel. Players choose to place bets on either a single number or a range of numbers, by placing chips on the corresponding location on the roulette table. To determine the winning numbers, a croupier spins the wheel in one direction, and then spins a ball in the opposite direction around a tilted circular track on the circumference of the wheel. The ball eventually loses momentum and falls onto the wheel and into one of thirty-seven coloured and numbered pockets on the wheel.

A summary of bet types, probabilities of success and payout ratios is given in Table 5.1. The payout ratio describes the amount of money returned to the player in the case of a winning bet, in addition to the amount they bet. For example, a payout ratio of 35:1 on a winning bet of $1 would return $36. Because the payout ratio directly corresponds to the amount of money a winner will receive, we select this as the basis for agents' calculation of incentive in the experiments in this chapter. The left-hand value of the payout ratio is divided by 35 to produce a normalised incentive value in the range [0, 1].

McClelland and Watson [4] studied individuals with three different motive profiles in human roulette experiments:

- **The nAch profile**: with 'high' achievement motivation and 'low' affiliation and power motivation,

- **The nAff profile**: with 'high' affiliation motivation and 'low' achievement and power motivation,

- **The nPow profile**: with 'high' power motivation and 'low' affiliation and achievement motivation.

Table 5.1 European roulette bets, probability of success, payout ratio and incentive assigned in the experiment in this chapter

Goal	Bet type	Numbers covered	Probability of success	Payout ratio	Incentive
G^1	Straight-up	1	1/37 (0.03)	35:1	1.00
G^2	Split-bet	2	2/37 (0.05)	17:1	0.49
G^3	Street-bet	3	3/37 (0.08)	11:1	0.31
G^4	Corner	4	4/37 (0.11)	8:1	0.23
G^5	Line-bet	6	6/37 (0.16)	5:1	0.14
G^6	Column	12	12/37 (0.32)	2:1	0.06
G^7	Dozen	12	12/37 (0.32)	2:1	0.06
G^8	Red or Black	18	18/37 (0.49)	1:1	0.03
G^9	Even or Odd	18	18/37 (0.49)	1:1	0.03
G^{10}	Low or High	18	18/37 (0.49)	1:1	0.03

We model these profiles computationally and use the experimental results reported by McClelland and Watson [4] as a benchmark for validating our models of achievement, affiliation and power motivation.

5.2.1 Motive Profiles for Roulette

We use the motivated rule-based agent algorithm in Algorithm 3.1 to create nAchAgents, nAffAgents and nPowAgents. The specific settings used are as follows:

- Line 1: Motivation is modelled using Eq. 2.9 with parameters in Table 2.5;

- Line 3: In this study we do not model sensors explicitly;

- Line 4: In the roulette game for artificial agents a computer-controlled agent can select from 10 condition-goal-behaviour tuples. Once again $C =$ true for all tuples in this simulation. There are 10 possible goals as shown in Table 5.1. A fixed betting amount was assumed for each bet, and incentive calculated as discussed above and shown in the last column of Table 5.1. Incentive is thus inversely proportional to probability of success, although following a different model to that proposed by Atkinson.

- Line 5: Goal selection is modelled using Eq. 2.15.

Comparison of the roulette and ring-toss experiments demonstrates the variation that is possible in incentive values or probabilities of success from one problem domain to the next. Humans deal with this smoothly. Individuals evaluate their current set of goals based on a relative ordering of success probabilities or incentives for the problems at hand. In real life, for example, individuals high in affiliation motivation tend to choose not to gamble at all [2]. When in a contrived,

Table 5.2 Motivation parameters and their experimental values or value ranges in the roulette experiment

Parameter	nAchAgents	nAffAgents	nPowAgents
S_{ach}	2.0	1.0	1.0
S_{aff}	1.0	2.0	1.0
S_{pow}	1.0	1.0	2.0
M_{ach}^+	[0.5, 0.6]	[0.5, 0.6]	[0.5, 0.6]
M_{ach}^-	[0.7, 0.8]	[0.7, 0.8]	[0.7, 0.8]
M_{aff}^+	[0.2, 0.3]	[0.2, 0.3]	[0.2, 0.3]
M_{aff}^-	[–0.2, –0.1]	[–0.2, –0.1]	[–0.2, –0.1]
M_{pow}^+	[0.8, 0.9]	[0.8, 0.9]	[0.8, 0.9]
M_{pow}^-	[1.0, 1.1]	[1.0, 1.1]	[1.0, 1.1]
ρ_{ach}^+	[0, 20]	[10, 90]	[0, 50]
ρ_{ach}^-	[0, 30]	[10, 80]	[0, 20]
ρ_{aff}^+	[30, 60]	[0, 100]	[70, 80]
ρ_{aff}^-	[60, 90]	[10, 80]	[80, 100]
ρ_{pow}^+	[10, 20]	[70, 90]	[50, 60]
ρ_{pow}^-	[0, 20]	[80, 100]	[0, 10]

S_{ach}, S_{aff}, S_{pow}, M_{ach}^+, M_{ach}^-, M_{aff}^+, M_{aff}^-, M_{pow}^+ and M_{pow}^- were chosen based on the psychological literature. ρ values were chosen experimentally

experimental situation, however, fewer goals are available, so these individuals must re-evaluate their goals relative to those that are currently available. The ability of humans to adapt their motives to different goal sets is reflected in the need to consider the motivation parameters for each problem. We do this here to create three motivated agent subtypes. The three subtypes are:

- **nAchAgents** have a higher value for S_{ach} and lower values for S_{aff} and S_{pow},

- **nAffAgents** have a higher value for S_{aff} and lower values for S_{ach} and S_{pow},

- **nPowAgents** have a higher value for S_{pow} and lower values for S_{ach} and S_{aff}.

The values of S_{ach}, S_{aff} and S_{pow} are fixed to create the distinct subtypes. For each subtype, one of the three values is higher than the other two. This assumes that there is a single dominant motive. Likewise, the ranges for M_{ach}^+, M_{ach}^-, M_{aff}^+, M_{aff}^-, M_{pow}^+ and M_{pow}^- are selected to conform to the various motivational subtypes. Specifically, they are selected so the local maxima of the motivation curve fall over mutually exclusive low, moderate and high incentive values, according to Table 5.1. The parameter ranges used are shown in Table 5.2.

Once again, the literature does not provide us with a way to select ranges for the various ρ values. In addition, an exhaustive search of the nature performed in Chap. 4 to find suitable upper and lower bounds for these parameters is now infeasible due to the size of the search space. However, we can approximate this search using a genetic algorithm with chromosomes encoded as follows:

$[lower(\rho_{ach}^+), \; upper(\rho_{ach}^+), \; lower(\rho_{ach}^-), \; upper(\rho_{ach}^-), \; lower(\rho_{aff}^+), \; upper(\rho_{aff}^+), \; lower(\rho_{aff}^-),$
$upper(\rho_{aff}^-), \; lower(\rho_{pow}^+), \; upper(\rho_{pow}^+), \; lower(\rho_{pow}^-), \; upper(\rho_{pow}^-)]$

Each gene can take a value between 0 and 100, following the schema used in Chap. 4 (see Table 4.4). A population of one thousand such chromosomes was initialised. Each generation, 250 pairs of parents were selected at random with replacement from the population. Crossover and mutation operators were applied to produce children. Two-point crossover was used, with the point chosen at random to occur at one of the locations marked with a ∥ below to ensure the integrity of the resulting child.

$[lower(\rho_{ach}^+), \; upper(\rho_{ach}^+) \parallel lower(\rho_{ach}^-), \; upper(\rho_{ach}^-) \parallel lower(\rho_{aff}^+), \; upper(\rho_{aff}^+) \parallel lower(\rho_{aff}^-),$
$upper(\rho_{aff}^-) \parallel lower(\rho_{pow}^+), \; upper(\rho_{pow}^+) \parallel lower(\rho_{pow}^-), \; upper(\rho_{pow}^-)]$

After crossover, a maximum of one randomly chosen gene could mutate with a probability of 0.01. With probability 0.005 the value of the gene increased, and likewise with probability 0.005 the value decreased.

The fitness of a chromosome was determined by randomly initialising a thousand agents of a given motivation type with S and M parameters as in Table 5.2 and ρ parameters in the ranges determined by the chromosome. The agents were then permitted to place one roulette bet, and the bet was recorded. The proportion of agents making each type of bet was compared against the proportion of humans in the McClelland and Watson [4] experiment placing the same bet. Fitness was computed to be proportional with the similarity between agent and human performance. The thousand fittest chromosomes were chosen for the next generation. After 300 generations the fittest chromosome discovered so far was used to inform the parameter ranges in Table 5.2.

5.2.2 Motivated Rule-Based Agents Playing Roulette

With parameter ranges identified, the human experiment can now be repeated with artificial agents. In this experiment, a thousand of each subtype of motivated agent was created using the parameter values in Table 5.2. Each simulated agents was allowed to select one bet. That is, each agent made one pass through the intelligence loop in Algorithm 3.1. The bet selected was recorded for each agent and the results collated. These results for the agents were then compared to experimental results with humans reported by McClelland and Watson [4].

The raw results for the number of bets of each type made by agent of each type are visualized in Fig. 5.1. Table 5.3 summarizes the salient statistics for these agents, including the average bet incentive selected, standard deviation and 95 % confidence interval. Table 5.3 indicates that there is a statistically significant difference in the behaviour of the three types of agents in terms of the average bet incentive selected.

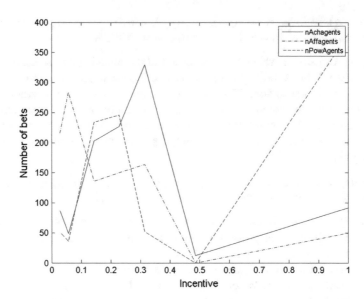

Fig. 5.1 Raw experimental results when 3,000 agents with three different types of motive profile can select roulette betting goals

Table 5.3 Summary of salient statistics for Fig. 5.2

	nAffAgents	nAchAgents	nPowAgents
Average bet incentive selected	0.178	0.287	0.491
Standard deviation	0.215	0.247	0.404
95 % confidence interval	±0.013	±0.015	±0.025

The 95 % confidence interval for the average bet incentive selected by nPowAgents is 0.466–0.516. Figure 5.1 shows that a majority of these agents select straight-up bets (G^1) or corner bets (G^4). Straight-up bets have the highest potential winnings but the least likelihood of success. In contrast, nAffAgents show a preference for low incentive bets, with an average bet incentive of just 0.165–0.191. Figure 5.1 indicates that the most preferred goals by nAffAgents are G^6–G^{10}. This result supports the avoidance of public competition seen in human participants. nAchAgents have an average incentive preference between that of nAffAgents and nPowAgents (0.272–0.302). Because these confidence intervals do not intersect, this result confirms that the behaviour of the three types of agents is statistically different at the 95 % confidence level.

As with human experiments, the results for nAchAgents require some interpretation. The most preferred bet by nAchAgents is the street-bet (G^3). The probability of success of a corner-bet is low in an absolute sense $(P^s(G^3) = 0.08)$, but can be thought of as intermediate in the roulette setting where both higher and lower odds bets are possible.

5.2.3 Comparison to Humans

Table 5.4 compares the experimental results for agents and humans, using the human results reported by McClelland and Watson [4] (p. 133). Their results are recorded only as a chart, the values of which we have estimated for the comparison in Table 5.4. McClelland and Watson [4] had 31 participants in their experiment: three affiliation-motivated, 13 achievement-motivated and 15 power-motivated. Each participant had 10 bets. McClelland and Watson [4] collated their results for each motive profile in a series of four brackets corresponding to 'moderate', 'low', 'very low' and 'extremely low' probabilities of success. These brackets are shown in the first column of Table 5.4. When multiplied by the three motive profiles (nAff, nAch and nPow), this gives a total of twelve experimental categories. Table 5.4 collates our agent results into these twelve categories and for each category shows the z-value (lowest entry) comparing the results for the agents (middle entry) with the results for humans (top entry) at the 95 % confidence interval. The critical z-value for a two-tailed z-test of two proportions at the 95 % confidence level is ± 1.96. Table 5.4 shows that our agents produce statistically similar results to those of human studies in all twelve categories. We note that the use of a genetic search algorithm to choose the motivation parameter values has improved this result compared to previous work with manually chosen parameters [5].

The next section demonstrates the same agents playing a third game—the prisoners' dilemma.

Table 5.4 Comparison of agents and humans in twelve motivation categories

Bracket	nAff Humans nAffAgents Z-Value	nAch Humans nAchAgents Z-Value	nPow Humans nPowAgents Z-Value
Extremely low probability of success: G^1–G^2	10.0 % 5.0 % −1.221	11.0 % 10.4 % −0.210	39.0 % 38.1 % −0.212
Very low probability of success: G^3–G^4	30.0 % 31.4 % 0.163	51.0 % 55.7 % 1.014	29.0 % 29.8 % 0.20
Low probability of success: G^5–G^7	37.0 % 42.0 % 0.547	26.0 % 25.2 % −0.197	27.0 % 27.0 % 0.00
Moderate probability of success: G^8–G^{10}	23.0 % 21.6 % −0.184	12.0 % 8.7 % −1.232	5.0 % 5.1 % 0.052

The percentage of humans and agents selecting goals in each category is shown. The critical z-value for this test is ± 1.96 at the 95 % confidence level

5.3   The Prisoners' Dilemma Game

The prisoners' dilemma (PD) [6] is one of the most widely studied problems in game theory literature. The PD is a two-player social dilemma game that gives specific meaning to the abstract game definition in Table 3.1 in Chap. 3. As in Chap. 3, each player in a PD game has two choices: cooperate, B^C, or defect, B^D. The payoff (gain or loss) for each player depends on the choices made by both players. A number of variations of PD games exist [3] in which players compete or cooperate to achieve a gain or avoid a loss. A classic example of the prisoners' dilemma in the loss domain is as follows:

Two suspects are arrested by the police. The police have insufficient evidence for a conviction, so they separate the prisoners and visit each of them to offer the same deal. If one testifies for the prosecution against the other (defects) and the other remains silent (cooperates), the defector goes free and the silent accomplice receives the full 10 year sentence. If both remain silent, both prisoners are sentenced to six months in jail for a minor charge. If each betrays the other, each receives a five year sentence. Each prisoner must choose to either defect or cooperate. How should the prisoners act?

Variations of the PD game have been studied in a range of fields as the generalised PD game has implications for studies in areas such as politics, science, economics, law and psychology. In psychology, variations of the PD game have been employed in conflict research [7] and uncertainty research, among others. Variations include one-shot games, iterative games, games in which players can communicate before making a decision, games where communication is forbidden, games in which players aim to achieve a gain and games in which players aim to avoid a loss.

In one-shot (static) games, players are allowed only one chance to make a decision. In iterative (dynamic) games players make a number of decisions and learn the outcome of their choice after each decision. Psychology experiments have shown that the differences in cognition and behaviour are most pronounced for one-shot decisions with relatively less threatening payoff matrices. For second and subsequent iterations, the game becomes more threatening and the temptation to defect and the fear of the sucker's payoff increase. Over time, players become more defensive and motivational differences diminish. In addition, in iterative scenarios, the players' experiences in the opening rounds strongly influence subsequent cooperation or defection [7]. One-shot games are the subject of this chapter. Iterative games and learning agents are studied in Part III.

Game-theoretic models are increasingly used in the artificial intelligence domain, especially to model strategic interactions between agents in multi-agent systems. These approaches attempt to mimic player behaviour using computer

Table 5.5 Payoff and corresponding explicit incentive in four prisoners' dilemma games

	Goals	Payoff (jail time) as 'years lost'	Explicit incentive
Game 1	G^T	$T = 0.0$	1.000
	G^R	$R = -0.5$	0.938
	G^P	$P = -5.0$	0.375
	G^S	$S = -8.0$	0.000
Game 2	G^T	$T = 0.0$	1.000
	G^R	$R = -0.5$	0.944
	G^P	$P = -5.0$	0.444
	G^S	$S = -9.0$	0.000
Game 3	G^T	$T = 0.0$	1.000
	G^R	$R = -0.5$	0.950
	G^P	$P = -5.0$	0.500
	G^S	$S = -10.0$	0.000
Game 4	G^T	$T = 0.0$	1.000
	G^R	$R = -0.5$	0.956
	G^P	$P = -5.0$	0.546
	G^S	$S = -11.0$	0.000

Game 1 is the least threatening. Game 4 is the most threatening, having the longest jail time

simulations that allow game-theoretic approaches to scale to large and complex systems. Various factors, such as an agent's expectancy of other agents' behaviour, may be captured in these simulations, but generally the agent models are based on the rationality assumption that agents compete to maximize their payoffs. More recently, the need for incorporating richer agent models that can capture cognitive processes has been highlighted [5]. The agent models in this chapter are developed with this aim in mind. Specifically, we are interested in creating agents with diverse decision-making characteristics that are similar to those of humans playing the same game. For simplicity, we limit our experiment to decision making in response to the payoff matrix, without taking into account other agents' expected behaviour.

In the remainder of this chapter, the four game outcomes (B^C, B^C), (B^D, B^D), (B^C, B^D) and (B^D, B^C) are thought of as implying four goals. In Table 5.5 these goals are labelled G^T, G^R, G^P and G^S. The agent's behaviour to cooperate or defect is determined by the goal they choose, which in turn depends on their individual motives.

Four different games are defined in Table 5.5 with increasingly threatening payoff matrices. Game 1 has the least threatening payoff matrix (a maximum of eight years in jail). Game 4 has the most threatening payoff matrix with a maximum of eleven years in jail. Payoff values V in these games are defined in the loss domain. However, incentive must be in the range [0, 1] to apply a motivation function. We calculate explicit incentive such that it is proportional to payoff as follows:

$$I^{s}(G) = \frac{V^{\max} - V}{V^{\max}} \qquad\qquad (5.1)$$

In each game the highest payoff is assigned an incentive of one. The lowest payoff is assigned an incentive of zero. Other incentive values are then computed proportionally to the original payoff.

5.3.1   Motivated Rule-Based Agents and the Prisoners' Dilemma

In this section we use the motivated rule-based agent algorithm in Algorithm 3.1 to create nAchAgents, nAffAgents and nPowAgents agents with the same parameters as those described in Sect. 5.2.1 (Table 5.2). We are thus investigating how well our agents tuned to play roulette can reproduce some of the results from human experiments with the prisoners' dilemma game.

The charts in Fig. 5.2 show the raw results from all four of the prisoners' dilemma games. Each chart shows the number of each type of agent selecting each type of goal: G^{T}, G^{R}, G^{P} or G^{S}. It should be noted that, of the four goals, G^{T} and G^{P} are both achieved by choosing to defect. Likewise, G^{R} and G^{S} are both achieved by choosing to cooperate. Figure 5.2 thus not only shows the behaviour chosen by each type of agent, but also the motivation for selecting that behaviour. We see that there is some preference for cooperation by nPowAgents and nAchAgents in Game 1, but this is replaced by a tendency towards the defensive G^{P} goal in the more threatening games. Table 5.6 shows the percentage of agents choosing to defect in each game. We see that, as the payoff matrix of the game becomes increasingly threatening, the percentage of agents choosing to defect increases. This is because the incentive of the G^{P} goal increases, making it more appealing for these agents.

5.3.2   Comparison to Humans

According to traditional game theory literature, two rational players should not cooperate when playing the PD game. However, in practice, in various human experiments the proportion of cooperative choices has ranged from 10 to 90 % [1]. Li et al. [3], for example, conducted a human experiment with an identical payoff matrix to that for our Game 3. They categorized human responses into two brackets corresponding to decisions to cooperate or defect, as shown in the first column of Table 5.7. Li et al. [3] reported 22 out of 62 human participants (35.5 %) choosing to cooperate. In our Game 3 a total of 1,424 out of the 3,000 agents (47.5 %) chose to cooperate. A two-tailed z-test of the two proportions results in z-values

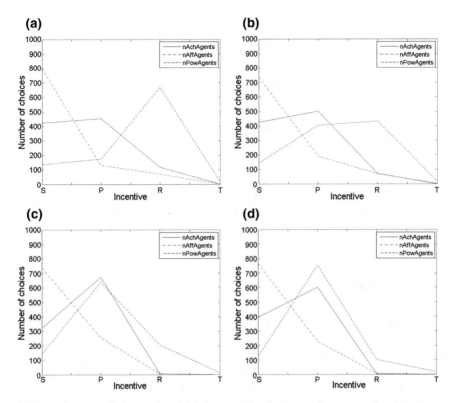

Fig. 5.2 Raw experimental results when 3,000 agents with three different types of motive profiles play prisoners' dilemma games: **a** Game 1 has the least threatening payoff matrix, followed by **b** Game 2 **c** Game 3 has a more threatening payoff than Game 2 and **d** Game 4 has the most threatening payoff matrix of the four games

Table 5.6 Percentage of agents choosing to defect for the four games in Table 5.5

	→ Increasing threat →			
	Game 1	Game 2	Game 3	Game 4
Percentage of agents choosing to defect	26.1	37.3	52.5	53.5

of ± 1.873 as shown in Table 5.7. These values are less than the critical z-value of ± 1.96 at the 95 % confidence level, confirming that the actions selected by humans and our agents are statistically similar at the 95 % confidence level.

The games we have chosen to examine in this section were selected because they have a similar incentive structure to that of the roulette experiment in Sect. 5.1, and have the same incentive structure to that of human experiments. As a result, it makes sense that the motivation parameter values developed in the roulette game should give us some sensible behavioural results in the PD game. However, it is also possible to envisage PD games with different payoff structures for which the

Table 5.7 Comparison of results from one choice each by 62 humans and one choice each by 3,000 agents ($1,000 \times$ nAffAgents, $1,000 \times$ nAchAgents, $1,000 \times$ nPowAgents)

Bracket	Humans Agents Z-Value
Cooperate: G^R, G^S	35.5 %
	47.5 %
	1.873
Defect: G^P, G^T	64.5 %
	52.5 %
	−1.873

Human results from Li et al. [3] are broken into two brackets. A two-tailed z-test of proportions is used to compare the results. The critical z-value is ± 1.96 at the 95 % confidence level

agents will have different behaviours. To this extent, the challenge of understanding how people interpret the payoff structures of these games through the lens of motivation remains open.

5.4   Conclusion

This chapter has shown how the computational models of motivation presented in Chap. 2 can be combined in an artificial motive profile and embedded in an agent framework. The chapter has introduced three types of motivated rule-based agents that use profiles of achievement, affiliation and power motivation. We call these agents nAchAgents, nAffAgents and nPowAgents.

Two experiments were presented that demonstrate the behaviour of artificial agents using the new models. These experiments are compared to similar real-world experiments with human participants to demonstrate the similarities between the behaviour of agents using the proposed new models and the observed behaviour of humans with corresponding motive profiles. Results permit us to conclude that:

- Agents with different motive profiles select their goals differently.

- In a risk-taking scenario achievement-motivated agents select goals of intermediate-high difficulty, power-motivated agents select high-incentive goals and affiliation-motivated agents avoid public completion by selecting low-incentive goals.

- When different games have similar incentive structures, it is possible for agents with motivation parameter values tuned for one game to play a second game and exhibit human-like behaviour.

The correlation between the results in this chapter and experimentally observed human behaviour confirm that computational models of achievement, affiliation and power motivation can offer a realistic starting point for building agents with motive profiles incorporating these motivations. Part III of this book will examine a number of game scenarios using such agents. In particular, it will extend the work in this chapter by studying agents involved in strategic interactions represented by a range of different two-player social dilemma games including the prisoners' dilemma game, the leader game, the chicken game and the battle of the sexes game. It will also extend the work in this chapter by permitting agents to make multiple decisions and learn over time.

References

1. R. Boyle, P. Bonacich, The development of trust and mistrust in mixed-motive games. Sociometry **33**, 123–139 (1970)
2. J. Heckhausen, H. Heckhausen, *Motivation and Action* (Cambridge University Press, New York, NY, 2010)
3. S. Li, Z.-J. Wang, Y.-M. Li, Is there a violoation of Savage's sure thing principle in the prisoner's dilemma game? Adapt. Behav. **18**, 377–385 (2010)
4. J. McClelland, R.I. Watson, Power motivation and risk-taking behaviour. J. Pers. **41**, 121–139 (1973)
5. K. Merrick, K. Shafi, Achievement, affiliation and power: motive profiles for artificial agents. Adapt Behav. **19**, 40–62 (2011)
6. W. Poundstone, *Prisoner's Dilemma* (Doubleday, New York, NY, 1992)
7. K.W. Terhune, Motives, situation and interpersonal conflict within prisoner's dilemma. J. Pers. Soc. Psychol. Monogr. Suppl. **8**, 1–24 (1968)

Part III
Game Scenarios for Motivated Agents

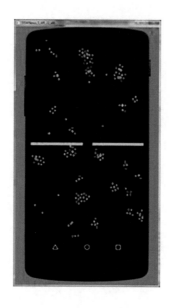

Chapter 6
Enemies

Part III of this book describes a range of in-game scenarios or mini-games that are appropriate for different types of motivated agents. Part III is broken into three chapters, corresponding to scenarios for three common non-player character types: enemies (this chapter), partner characters (Chap. 7) and support characters (Chap. 8). This chapter considers two kinds of scenarios in which motivated agents are used to augment the behaviour of enemies: the first for motivated rule-based agents and the second for motivated learning agents. Theoretical and empirical analyses are performed and two applications presented for a single-player shooter game and a turn-based strategy game. We show how theoretical and empirical results can be used to predict the behaviour of agents in an application.

6.1 Types of Non-player Characters and Their Roles

Non-player characters can be classified in three main categories: enemies, partners and support characters [7]. In this book we are particularly interested in the role that non-player characters take in the competitive aspects of gaming. We saw in Chap. 1 that motivation and competitive behaviour are closely linked, with motivation influencing factors such as desire to win, willingness to take risks and willingness to engage in conflict or cooperation.

Enemies are a key character type for implementing the competitive aspects of gaming. The most aggressive enemies oppose human players in a pseudo-physical sense by attacking the virtual life force of the human player with weapons or magic. A broader concept of enemies can also include opponents and competitors. Opponents may provide strategic opposition, while competitors may vie for the same goal as the player, without specifically attacking the player's virtual life force. Enemies, opponents and competitors, together, form a critical character group in a wide range of game genres from action games to strategy or sport games.

© Springer International Publishing AG 2016
K.E. Merrick, *Computational Models of Motivation for Game-Playing Agents*,
DOI 10.1007/978-3-319-33459-2_6

From a competitive perspective, partners take the opposite role of enemies and attempt to protect human players with whom they are allied. Partner characters might also perform scripted non-combat roles to support their human ally. In some games, partner characters may be taught to perform certain behaviours by players.

Finally, support characters are the merchants, tradesmen, guards, innkeepers and so on who support the story line of the game. Support characters are not necessarily involved directly in the competitive aspects of the game. However, they support competition indirectly by offering quests, advice, goods for sale or training to the player.

Of the three categories described above, artificial intelligence research for computer-controlled enemies is arguably the most highly developed. Enemies must be strong enough to be challenging, but have sufficient weaknesses to permit them to be beaten by human players. It is not the aim of motivated agent technologies to compete with traditional approaches to the design of enemies and competitors, but rather to provide a novel tool for introducing greater diversity into their ranks.

To demonstrate this, this chapter considers a number of abstract and concrete scenarios for the algorithms presented in Chap. 3. Section 6.2 presents a concrete scenario for the application of motivated rule-based agents. It works through a methodology for a simple application of Algorithm 3.2 in a custom implementation of a real game—Greg Kuperberg's 1982 classic *Paratrooper* [6]. Sections 6.3 and 6.4 consider a spectrum of abstract and concrete scenarios for motivated learning agents. Each of these sections presents an abstract scenario modelled by a mixed-motive game. Section 6.4 considers an application in a scenario from a turn-based strategy game.

6.2 *Paratrooper*

Paratrooper [6] is a single-player shooter game developed in the early 1980s. As a stand-alone game, *Paratrooper* is simple by today's standards. However, it is not unusual to find similar scenarios embedded as mini-games in more complex, contemporary massively multiuser online games. *Paratrooper* is a useful game to study in the context of enemy non-player characters (NPCs) because all of its NPCs are enemies. It has a number of different types of enemies and these characters form an implicit hierarchy. The visible NPCs in this game are helicopters, paratroopers and bombers. The game controller influences when helicopters and bombers enter the screen. Helicopters are responsible for dropping the paratroopers.

The game is suitable for the introduction of motivated agents because each NPC has a clear spectrum of similar goals, which we identified in Sect. 3.2.1 as a requirement for motivated rule-based agents. Bombers, for example, may have a set of goals for different flight altitudes. Helicopters may have a set of goals for dropping paratroopers at different distances from the gun. Paratroopers themselves may have a set of goals for opening their parachutes at different heights. This is the aspect of the game we investigate in this section.

A player of *Paratrooper* controls a gun turret at the bottom centre of the screen. When the player pushes the left or right arrow keys, the turret swivels in the appropriate direction until another arrow key is pressed or the gun is fired. The gun may fire multiple shots at once, but only as quickly as the player can press the fire key (space bar). When the gun is fired it will stop in its current position.

Scoring for the game starts at zero. Each shot fired costs one point, although the minimum score is zero. Hitting a bomber or helicopter earns the player 10 points, hitting an enemy paratrooper earns the player five points, and hitting a bomb, 30 points. To maximise their score a player should be both accurate and conservative with ammunition.

The behaviour of NPCs in the original *Paratrooper* is rule-based and probabilistic. The helicopters enter the screen at three fixed heights. The helicopters flying right to left take the highest entry point, while the helicopters flying left to right enter below. The helicopters have the ability to drop paratroopers, although they do not always invoke this ability. Any helicopter can drop up to three troopers. The paratroopers, once released from the helicopter, are in free fall until they release their parachute. At the point they release their parachute their descent speed decreases.

After approximately 15–20 helicopters have been destroyed, the first stage of the game ends. The second stage of the game involves the bombers. Bombers always enter the screen at the maximum height. The game usually spawns two to four bombers from the left of the screen, with the possibility of two to four from the right after the first wave. The bombers have the ability to drop bombs, but do not always invoke this ability.

The game ends when the player's gun turret is hit by a bomb; when a single paratrooper lands directly on the turret; or when four paratroopers safely land on either the left or the right of the turret (once this happens, they are able to build a human pyramid, climb up to the turret and blow it up).

A screenshot from our custom implementation of the *Paratrooper* game [13] is shown in Fig. 6.1. Our implementation was done in Java. It uses the game logic described above, but is flexible enough to incorporate artificial agents in the various NPCs, including the paratroopers, helicopters, bombers and the game controller itself.

The next section describes an implementation of motivated rule-based agents in the game's main antagonists—the paratroopers.

6.2.1 Motivated Rule-Based Agents as Paratroopers

We saw in the text above that the game controller, helicopters and paratroopers form one branch of the hierarchy of NPCs in *Paratrooper*. The game controller and bombers form another. In other words, the game controller can influence both helicopters and bombers, but the helicopters and bombers in the original form of this game do not influence each other.

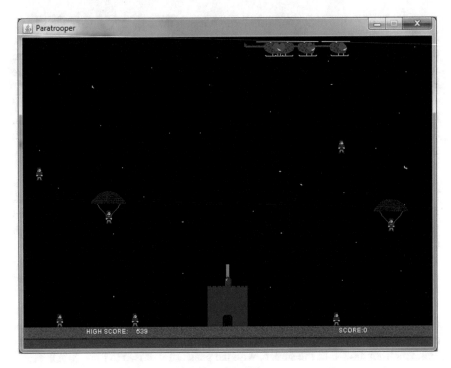

Fig. 6.1 A custom implementation of *Paratrooper* [6]

This section describes how we can embed motivated rule-based agents with probabilistic goal selection (Algorithm 3.2) in characters at the bottom of this hierarchy—the paratroopers. We consider this the first phase of introducing computational motivation to a game. In Chap. 9 we will consider a second phase that introduces motivation higher in the NPC hierarchy.

The specific settings we used for Algorithm 3.2 are:

- Line 1: Motivation is modelled using the 'dominant motive only' method (Eq. 2.10) with the parameters in Table 6.1;

- Line 3: Paratroopers can sense their height above the ground and whether their parachute is open or not. That is, $S(W_t) = \langle h, o \rangle$. Height h above the ground can take one of 50 discrete values, while o is binary. In the implementation studied here the sensed state is used only in the execution of

Table 6.1 Motivation parameters and their experimental values in the paratrooper experiment

Parameter	nAchAgents	nAffAgents	nPowAgents
S_{mot}	0.5	0.5	0.5
M_{mot}^{+}	0.25	0.1	0.6
M_{mot}^{-}	0.75	0.3	0.9
ρ_{mot}^{+}	20	20	20
ρ_{mot}^{-}	20	20	20

behaviour. Motivated decision making is governed by a set of goals described in the next bullet point;

- Line 4: There are 50 condition-goal-behaviour tuples. C = true for all tuples. That is, state is not considered for motivated decision making. There are 50 goals, G^1–G^{50}, for opening the parachute at one of 50 positions above the ground. Each corresponding behaviour B^g permits the paratrooper to free-fall to the specified altitude then open their chute on arrival at that height. $I^s(G^g)$ depends on g. Specifically, opening the chute closer to the ground has a higher incentive than opening the chute higher in the air. Opening the chute closer to the ground is riskier, because any error in judgement can result in the chute failing to open in time. However, it also means that the paratrooper completes their descent more quickly, potentially making it more difficult for them to be shot by the gun turret. The relationship between height above the ground and incentive is modelled by a linear function with 50 discrete heights mapped to incentive values between 0 and 1.

- Line 5: Motivation is computed using Eq. 2.10

- Line 6: A goal is selected probabilistically using Eq. 2.16

The effect of these settings is that paratroopers select a chute-opening altitude probabilistically from their set of goals as they exit their helicopter, according to their motive profile. Their behaviour is then to drop to the selected height and open their chute on arrival at that height.

6.2.2   Empirical Study: Motivated Paratroopers

Because the game deploys only a small number of paratroopers at a time, we analyse the expected behaviour of large numbers of paratroopers using a MATLAB simulation. We perform further in-game analysis in Chap. 9.

Using the parameter values in Table 6.1, 1,000 paratroopers were generated and the heights at which they opened their parachutes recorded. The results are shown in Fig. 6.2. We see that:

- The heights at which agents with different motives open their parachutes are distributed differently for agents with different motives. That is, motivation influences goal selection and behaviour;

- Power-motivated agents open their chutes closer to the ground. As we will see in Chap. 9, the resulting extended period of fast free fall gives them a slightly greater chance of reaching the ground and a greater chance of besting the player;

- Affiliation-motivated agents open their chutes higher in the sky;

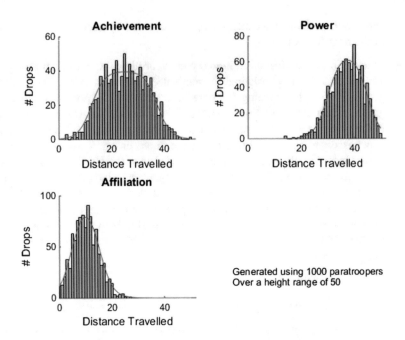

Fig. 6.2 Distribution of distance travelled before opening of the parachute for motivated rule-based agents with different dominant motives

- Achievement-motivated agents open their chutes at moderate heights. This represents a calculated risk in our scenario, in keeping with the characteristics of achievement-motivated agents presented in Table 1.2.

This brief analysis shows how the use of motivation permits us to develop characters with predictable diversity of behaviour, and how we can measure that diversity objectively. In contrast to the original game, in which all paratroopers operate by the same rules, with the introduction of motivation we now have a heterogeneous set of paratroopers. This means that we now have the option of introducing a higher level of motivated reasoning within the game. We will do this in Chap. 9, where we return to this game to study the evolution of motivated agents via the game controller.

6.2.3 Summary

The application above uses arguably the simplest implementation of motivated rule-based agents in this kind of a game scenario. There are thus a number of alternative ways that the agents could be implemented, including:

- Making the sensed state more complicated. For example, we could include variables such as the angle of the gun and the paratrooper's distance from the gun when dropped. This information could be included in the calculation of goal incentive. It would permit the agents to have a higher-resolution view of their environment and take more factors into account when computing motivation for each goal;

- Permitting the motivated paratroopers to make multiple decisions over time as they drop. This would permit them to change their decision regarding when to open the chute in response to changes in the state of the game. This of course requires greater use of computational resources.

The remainder of this chapter, however, is concerned with further understanding and predicting the characteristics of agents with different motives. To do this we will first return to the prisoners' dilemma.

6.3   The Prisoners' Dilemma Revisited

In Chap. 5 we saw that motivated agents can be designed to exhibit some of the characteristics of humans when playing certain games. One of those games was the prisoners' dilemma (PD). Chapter 5 examined a type of play in which agents were permitted only a single decision. In contrast, this chapter examines 'iterative' play, when agents can make multiple decisions and learn from previous decisions when making new ones. This is of course a useful trait for enemy NPCs.

To review, the PD game [12] derives its name from a hypothetical strategic interaction in which two people are arrested for involvement in a crime. They are held in separate cells and cannot communicate with each other. The police have insufficient evidence for a conviction unless at least one of the prisoners discloses certain incriminating information. Each prisoner has a choice between concealing information from the police (B^C) or disclosing it (B^D). If both conceal, both with be acquitted. The payoff for this outcome is the same for both players: $V^1 = V^2 = R$. If both disclose, both will be convicted and receive minor punishments: $V^1 = V^2 = P$. If only one prisoner discloses information he will be acquitted and, in addition, receive a reward for his information. In this case, the prisoner who conceals information will receive a heavy punishment. For example the payoff if A^1 discloses in the hope that A^2 conceals is $V^1 = T$. The payoff if A^2 conceals when A^1 discloses is $V^2 = S$. A^2 in this situation is sometimes referred to as the 'martyr' because he generates the highest payoff for the other player and the lowest payoff for himself.

If we express the text above formally, the PD game is a two-by-two mixed motive game with general form presented in Chap. 3, Table 3.1 and Eq. 3.12. The specific payoff structure of a PD game is $T > R > P > S$. In addition, the iterative version of the PD traditionally requires that $2R > T + S$.

The PD game can be used as a model for real-world social dilemmas such as arms races (see also Chap. 3), voluntary wage restraint, doping in sport [4], conservation of scarce resources, advertising and climate change action (also see [3] for a review). The PD game can also be used to model scenarios such as arms races in computer games. This scenario is the focus of this study.

6.3.1 Arms Races in Turn-Based Strategy Games

Arms races often occur in turn-based strategy (TBS) games like Sid Meier's *Civilisation* [8] series. In this game series, NPCs have a choice of pursuing a peaceful, scientific victory, or a war-like military victory against a human opponent. Mapped to a PD game, the B^D choice represents assignment of resources to build weapons and the B^C choice represents assignment of resources to other national priorities. A NPC that chooses B^D when the opposing player does not has an advantage because it possesses weapons that could potentially subdue the other. When both parties build weapons ((B^D, B^D) outcome), there is no such advantage, but both parties have put their resources into the development of weapons at the expense of other national priorities. The (B^C, B^C) outcome thus has a higher payoff as both parties can assign their resources elsewhere to increase the size of their cities and the happiness of their subjects. We see from this example that the B^C and B^D choices are high-level plans of play, not individual actions. For example, the B^D choice might be executed by a series of actions for building military units, fortifying cities and so on.

In the arms race described above, there is a pure strategy equilibrium point (B^D, B^D) from which neither player benefits from unilateral deviation, although both benefit from joint deviation. For this reason, classical game theory predicts that all rational players who correctly perceive the game will ultimately choose the (B^D, B^D) outcome when playing this particular social dilemma game. However, in practice that is often not the case [5]. Humans with different motives consider alternative strategies that avoid the (B^D, B^D) outcome. Likewise, in computer games, NPCs are programmed with rules to control when they offer a peace treaty and when they declare war. These rules might, for example, be based on the military strength of the opposing player. This section describes an alternative approach in which a motivated agent learns which decision to make based on its own motives and the choices made by its opponent.

6.3.2 Perception of Arms Races by Motivated Learning Agents

As intimated above, and indeed throughout this book, one of the effects of motivation is to cause a form of misperception of the scenario at hand. Agents with different motives 'misperceive' scenarios differently. In this section, we consider how the arms race described above is misperceived as a different game, depending on the motive profile of the agent. We do this by using the optimally motivating incentive (OMI) representation for motivation, and formally defining the matrix transformations that occur for a given value of Ω^j. We divide the transformations into six cases corresponding to perception caused by different strengths of power motivation, achievement motivation and affiliation motivation. We use the following mapping of OMI values to motivation types:

1. nPow(1) agents: strong power-motivated agents, $T > \Omega^j > 1/2(T+R)$

2. nPow(2) agents: weak power-motivated agents, $1/2(T+R) > \Omega^j > 1/2$ $(T+P)$

3. nAch(1) agents: achievement-motivated agents, $1/2(T+P) > \Omega^j > 1/2$ $(R+P)$

4. nAch(2) agents: achievement-motivated agents, $1/2(R+P) > \Omega^j > 1/2$ $(R+S)$

5. nAff(1) agents: weak affiliation-motivated agents, $1/2(R+S) > \Omega^j > 1/2$ $(P+S)$

6. nAff(2) agents: strong affiliation-motivated agents, $1/2(P+S) > \Omega^j > S$

This mapping implies that power-motivated individuals prefer 'high' incentives, such that the definition of 'high' is determined by the highest three payoff values (T, R and P in this case). Achievement-motivated individuals prefer 'moderate' incentives, with the definition of 'moderate' being governed by all of the payoff values. Affiliation-motivated individuals prefer 'low' incentives, defined by the lowest three payoff values. This mapping of motivation types to incentive values implies that 'high', 'moderate' and 'low' can change depending on the distribution of values T, R, S and P.

The division of OMIs into 'high', 'moderate' and 'low' ranges is informed by the literature on incentive-based motives discussed in Chap. 1, but represents a stronger mathematical assumption that has previously been made by psychologists. Specifically, while previous experiments with humans have provided evidence that individuals with different dominant motives prefer different goals, these experiments have not attempted to define precise value ranges for the incentives that will be preferred by different individuals. In artificial systems, such a precise mapping is critical for defining and evaluating different types of agents.

Because we represent motivation using an OMI here, we no longer need to perform a calculation such as that of Eq. 5.1 to normalise the range of incentive to [0, 1]. Rather, we can calculate the subjective incentive of a payoff value directly using Eq. 2.13. Suppose we have a PD game, $\mathbf{W}$, in the matrix form given in Eq. 3.12. Then, using Eq. 2.13 and the knowledge that $V^{max} = T$, we can construct a perceived game $\widehat{\mathbf{W}}$ as shown in Table 6.2. We can see from Table 6.2 that, because Ω^1 and Ω^2 may be different, the values perceived by each player may also be different. In fact, there are eight different games that can be perceived by motivated agents, depending on their OMI and the distribution of incentive values in the original game. The theory required to derive these transformations is presented in Appendix A.

Table 6.2 Perceived game $\widehat{W}$ and subjective incentives $\widehat{T}, \widehat{P}, \widehat{S}$ and $\widehat{R}$ assuming that T is the maximum explicit incentive value in the original game W

		Player 2	
		B^D	B^C
Player 1	B^D	$\widehat{P}^1 = T - \lvert P - \Omega^1 \rvert, \; \widehat{P}^2 = T - \lvert P - \Omega^2 \rvert$	$\widehat{T}^1 = T - \lvert T - \Omega^1 \rvert, \; \widehat{S}^2 = T - \lvert S - \Omega^2 \rvert$
	B^C	$\widehat{S}^1 = T - \lvert S - \Omega^1 \rvert, \; \widehat{T}^2 = T - \lvert T - \Omega^2 \rvert$	$\widehat{R}^1 = T - \lvert R - \Omega^1 \rvert, \; \widehat{R}^2 = T - \lvert R - \Omega^2 \rvert$

The eight perceived games are visualised in Fig. 6.3. Figure 6.3a shows that nPow(1) agents with the highest OMIs still perceive a PD game. However, in contrast, it is clear from the visualisation that the strategy of always playing B^D no longer dominates for nPow(2) agents (see Fig. 6.3b). An equilibrium can occur if both players choose the 'always play B^D' strategy or if both players choose the 'always play B^C' strategy. The equilibrium that occurs depends on the initial probabilities of $P^1(B_0 = B^C)$ and $P^2(B_0 = B^C)$. For fixed values of $\widehat{P}^1$ and $\widehat{S}^1$, the further $\widehat{R}^1$ is above $\widehat{T}^1$, the greater the range of initial values of $P^1(B_0 = B^C)$ and $P^2(B_0 = B^C)$ that will result in the equilibrium in which both players play B^C (graphically, imagine the intersection points of the $E^1(B^D)$ and $E^1(B^C)$ lines moving further to the left). The choice of values for T, R, P and S thus plays a role influencing the equilibrium that will occur over time as a result of learning.

Unlike power-motivated agents, achievement-motivated agents may perceive one of four different games, shown in Fig. 6.3c–f. The transformation of a game depends on the distribution of incentive values as well as the value of Ω^j. The games in Fig. 6.3d, e result if the values of T, R, P and S have a highly nonlinear distribution, with a much larger gap between T and R than between the other incentives. The games in Fig. 6.3c, f result from all other distributions, including linear distributions.

Figure 6.3g, h visualise the structure of the games perceived by affiliation-motivated agents. We see in Fig. 6.3g that nAff(2) agents with a very low OMI will perceive a game in which the 'always play B^C' strategy dominates the 'always play B^D' strategy. This means that nAff(2) agents will always converge on the (B^C, B^C) outcome over time, even though the (B^D, B^D) outcome has greater subjective incentive to both.

In summary, misperception as a result of motivation produces a number of different perceived games. Some power-motivated individuals will still perceive a game with the same payoff structure as the original PD game, but agents with other motives will perceive one of seven different games with different equilibrium points. Merrick and Shafi [10] used two population replicator dynamics to simulate the learning trajectories when pairs of players with the same motive profiles play each other. In contrast, this chapter will examine learning trajectories when a NPC with a given profile interacts with a simulated player who has a different motive profile.

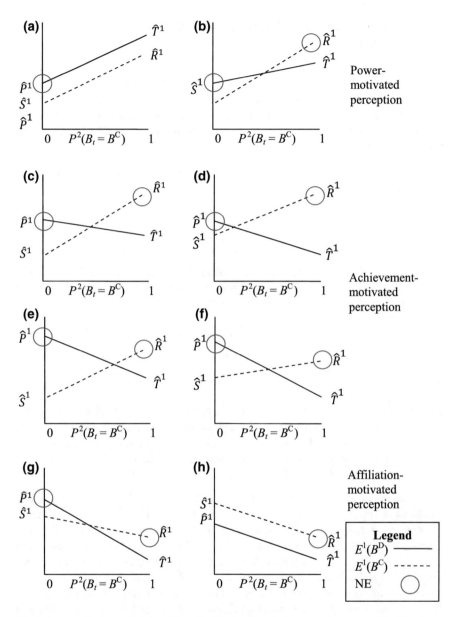

Fig. 6.3 Visualisation of the prisoners' dilemma game perceived by **a** nPow(1) agents (Theorem A.1); **b** nPow(2) agents (Theorem A.2); **c** and **d** nAch(1) agents (Theorem A.3); **e** and **f** nAch(2) agents (Theorem A.4); **g** nAff(1) agents (Theorem A.5). **h** nAff(2) agents (Theorem A.6). The Nash Equilibria (NE) are circled

6.3.3   Empirical Study: Motivated Learning
Agents in an Arms Race

This section presents experiments in which agents with different OMIs play the PD game. Because both artificial agents (representing NPCs) and human players can now have motives, we examine the experiments from the perspective that one agent represents an NPC and the other a player. The agents use Algorithm 3.4 with specific settings as follows:

- Line 1: **W** is defined as in Eq. 6.1. This game has a linear payoff structure in the gain domain:

$$W = \begin{bmatrix} 2 & 4 \\ 1 & 3 \end{bmatrix} \tag{6.1}$$

- Line 2: In each experiment the interacting agents have different OMIs. Specifically, in each of the experiments, two of the following three types of motivated agents play against each other:

$$\text{nPow(1) agents with } \Omega^j = 3.9$$
$$\text{nAch(1) agents with } \Omega^j = 2.6$$
$$\text{nAff(2) agents with } \Omega^j = 1.1$$

According to the analysis in Sect. 6.3.2, agents with these OMIs will have the most significant differences in perception of the original game. The initial probabilities $P^1(B_0 = B^C)$ and $P^2(B_0 = B^C)$ also required in line 2 to initialise the agents are selected randomly from a uniform distribution. $P^1(B_0 = B^D) = 1 - P^1(B_0 = B^C)$ and $P^2(B_0 = B^D) = 1 - P^2(B_0 = B^C)$. $\alpha = 0.001$.

We examine the behaviour of agents interacting and learning during 3,000 iterations of each game. The results in this section are presented visually in terms of the change in probabilities of agents choosing B^C (and by implication, B^D). The horizontal axis on each chart measures $P^1(B_t = B^C)$ for a NPC and the vertical axis measures $P^2(B_t = B^C)$ for a player. Thus, a point in the upper right corner of each graph is indicative that both agents are choosing B^C. The other corners also have a prescribed meaning, as shown in Fig. 6.4. Points closer to the centre of the chart are indicative of a more or less equal preference for B^C and B^D. $P^1(B_0 = B^C)$ and $P^2(B_0 = B^C)$ are initialised randomly, so the learning trajectories start in the middle region of the charts and proceed to the corners.

The lines in Fig. 6.4 trace the learned values of $P^1(B_t = B^C)$ and $P^2(B_t = B^C)$ over time. These lines are the learning trajectories of the NPC and the simulated

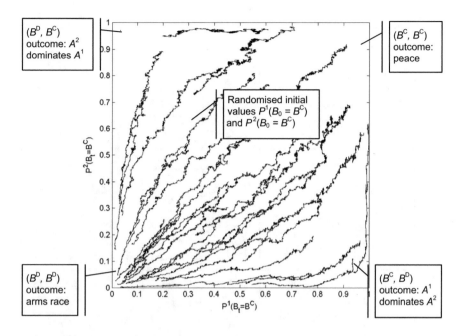

Fig. 6.4 Simulation of thirty pairs of agents playing 3,000 iterations of the prisoners' dilemma game (contextualised as an arms race). All are nPow(1) agents, but initial values $P^1(B_0 = B^C)$ and $P^2(B_0 = B^C)$ are randomized from a uniform distribution

player, with one line representing the trajectory for each NPC-player pair. In Fig. 6.4 all agents are nPow(1) agents and thus still perceive a PD game. We see that, because the perceived games are PD games, all agent pairs tend to converge on the (B^D, B^D) Nash equilibrium (NE) over time. In the context of our example, we can say that an arms race results. This outcome is the non-cooperative (i.e. competitive) outcome. Competitive behaviour is one of the characteristics of power-motivated individuals recognised by psychologists that we saw in Chap. 1 (see Table 1.2).

Figure 6.5 visualises the learning trajectories in nine cases, corresponding to three motive profiles for NPCs and three for players. When NPCs are nPow(1) agents (column 1 in Fig. 6.5), we see that they universally adopt the exploitative 'always play B^D' strategy. Power-motivated agents here conform to Bartle's 'killer type'. That is, they choose to build their military resources regardless of the type of opponent they face.

In contrast, when NPCs are nAch(1) agents (middle column in Fig. 6.5), we see that they adopt different strategies depending on the decisions made by their opponent. They learn to build their military when playing against nPow(1) players, but focus on the non-militarised outcome when playing against nAff(2) players. One feature to note is that the learning trajectories of nAch(1) agents do not proceed

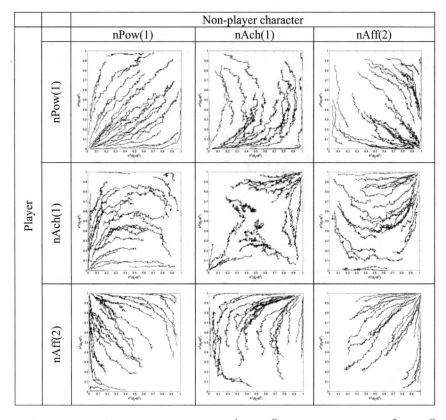

Fig. 6.5 Each subfigure shows the change in $P^1(B_t = B^C)$ (*horizontal axis*) and $P^2(B_t = B^C)$ (*vertical axis*) when 30 pairs of agents with different motives play 3,000 iterations of the prisoners' dilemma game

directly to the corners of the figures. Rather they curve one way and then back into the equilibrium point. In the case of nAch(1) NPCs opposing nPow(1) players, for example, this means that the nAch(1) agents initially increase their preference for a non-militarised strategy, but when they are consistently exploited by their nPow(1) opponents they begin to shift their strategy to build their military units. An arms race results.

Finally, the nAff(2) agents (last column in Fig. 6.5) universally choose a non-militarised strategy ('always choose B^C'), even though this sometimes means they are exploited. This is of course the cooperative outcome, and cooperative or collaborative behaviour is one of the recognised characteristics of affiliation-motivated individuals that we saw in Chap. 1 (Table 1.2).

We note that with a learning rate of $\alpha = 0.001$ it takes a few thousand iterations before behaviour is observed that converges on the game's theoretical equilibrium. This gradual adaptation can be advantageous for NPCs that require lifelong learning. Alternatively, speed of learning can be increased by increasing α. The effect of increasing α is to permit the agent to make larger adjustments to its values of $P^j(B_0 = B^C)$ [11].

Each tenfold increase in α speeds convergence roughly tenfold, meaning that 10 times fewer decisions are made before $P^1(B_t = B^C)$ and $P^2(B_t = B^C)$ stabilise. The trade-off is that the learning trajectory is less predictable. The agents can have large changes in $P^j(B_t = B^C)$ and these changes can be either increases or decreases. Thus, while the eventual equilibrium remains predictable, the trajectory taken to get there can be highly variable.

In terms of selecting a motivated learning agent to control a NPC in scenarios that can be modelled by a PD game, we can conclude that:

- nPow(1) agents will adapt to exploit (choose B^D) all opponents. They will exhibit characteristics of competitive behaviour.

- nAch(1) agents will adapt differently to different opponents, choosing B^D when exploited, but B^C when their opponent does likewise.

- nAff(2) agents will choose B^C against all opponents. They will exhibit characteristics of cooperative behaviour.

These simulations give us an initial understanding of the adaptive diversity that can be achieved using motivated learning agents to control an enemy NPC. There are clear differences in the emergent behaviour of agents with different motivations, and these differences manifest themselves when the agents interact. There is some correlation between the characteristics that psychologists associate with different motivation types and the types of behaviour exhibited by agents with different motives.

The next section considers how these characteristics emerge in a different abstract game modelling competitive exploration. This game, called the leader game, is still a two-by-two mixed motive game, like the PD, but the payoff structure is different. It thus has a different set of real-world (or virtual world) analogies.

6.4 The Leader Game

The leader game is so named as it is an analogy for real-world interactions such as those between pedestrians or drivers in traffic. For example, suppose two pedestrians wish to enter a turnstile. Each must decide whether to walk into the turnstile first (B^D) or concede right-of-way and wait for the other to walk in (B^C). If both pedestrians wait, then both will be delayed. The payoff for this outcome is the same for both players, $V^1 = V^2 = R$. If they both decide to walk first, a socially awkward situation results. The payoff to both players is again the same, $V^1 = V^2 = P$. If one decides to walk and the other waits, the 'leader' will be able to walk through unimpeded. The payoff for this outcome is T. The 'follower' will be able to walk through afterwards (payoff S). To summarise the payoff structure of the leader game we have $T > S > R > P$.

Leader games occur frequently in the context of computer games that require exploration of a game map. This is the case in many turn-based strategy (TBS) games, such as those from the *Civilisation* [8], *Alpha Centauri* [9] and *Heroes of Might and Magic* [15] series. Leader games may involve two opponents (as considered in this chapter) or multiple opposing parties (as considered in Chap. 9). As we will see in Chap. 9, the *n*-player case is merely a compound version of the two-player case.

In the TBS games listed above, the game map is initially invisible to the player except for a small starting area. Players must move their units (such as settlers or military units) off the edges of the known world to reveal more terrain. There are generally collectable resources to be discovered in huts (*Civilisation*) or supply pods (*Alpha Centauri*). It is thus desirable to be the first player to explore a given area (B^D), although traversal of an already explored area (B^C) can be beneficial for finding new places to expand one's empire. Players and NPCs must decide whether to compete to be the first to explore the terrain (the (B^D, B^D) outcome), or avoid conflict by letting the other lead the exploration (the (B^D, B^C) or (B^C, B^D) outcomes). The competition becomes particularly apparent during the mid-late stages of a TBS game, when players can see each other on the map. Alternatives to exploratory behaviour include terraforming or irrigating the terrain, building roads and so on. These activities add value, but not in the same immediate way as discovery of a new collectable resource. The (B^D, B^C) outcome thus has the highest payoff of $V^1 = T$ for the first mentioned player (the 'leader') and $V^2 = S$ for the second (the 'follower'). If both players explore the same part of the map at the same time (B^D), then they may need to battle each other for any resources uncovered. This battle usually results in the death or decimation of at least one of the player's units. This is thus the lowest payoff outcome for both $V^1 = V^2 = P$. However, if both players wait for the other to explore, the benefits of exploration cannot be collected. The payoff for both players will be R.

We saw in Chap. 1 that exploration and discovery have been considered in the context of both player types [2] and motivational subcomponents [17]. This application is thus interesting as a way to investigate the creation of NPCs with different preferences for exploration. We can follow the same process used in Sect. 6.3.2 and Appendix A to construct perceived versions of the leader game. For the leader game, we have the following definitions for motivated agents:

1. nPow(1) agents: strong power-motivated agents, $T > \Omega^j > 1/2(T+S)$

2. nPow(2) agents: weak power-motivated agents, $1/2(T+S) > \Omega^j > 1/2(T+R)$

3. nAch(1) agents: achievement-motivated agents, $1/2(T+R) > \Omega^j > 1/2(S+R)$

4. nAch(2) agents: achievement-motivated agents, $1/2(S+R) > \Omega^j > 1/2(S+P)$

5. nAff(1) agents: weak affiliation-motivated agents, $1/2(R+P) > \Omega^j > 1/2$
    $(R+P)$

6. nAff(2) agents: strong affiliation-motivated agents, $1/2(R+P) > \Omega^j > P$

The definitions have the same structure as those for agents playing the PD game in Sect. 6.3.2. Power-motivated agents have OMIs chosen based on the highest three payoff values; the OMIs for achievement-motivated are influenced by all four payoff values; and the OMIs for affiliation-motivated agents are influenced by the lowest three payoff values. The definitions differ slightly from those for agents playing the PD game, due to the difference in payoff structure of the leader game. However, the agents still perceive eight games in the leader scenario, which we will discuss in the next section.

6.4.1 Perception of Leader Games by Motivated Learning Agents

Figure 6.6 visualises the structure of the eight games that can be perceived when motivated agents engage in a scenario modelled by a leader game. nPow(1) agents with the highest OMI still perceive a leader game as shown in Fig. 6.6a. However, Fig. 6.6b shows that nPow(2) agents misperceive the game. Although there are still two equilibria, (B^C, B^D) and $(B^{\bar{C}}, B^D)$, the NPC prefers the latter of these when they have an nPow(2) profile.

Figure 6.6c, d visualise the structure of the games perceived by nAch(1) agents according to Theorem A.9. The game structure shown in Fig. 6.6d is perceived if the original game has a highly nonlinear payoff distribution. The game structure in Fig. 6.6c is perceived if the original game has a linear or close to linear payoff distribution. However, the strategy of always playing B^C dominates for all nAch(1) agents, regardless of the distribution of the payoff. Likewise, nAch(2) agents perceive the game in Fig. 6.6e when the original game has a highly nonlinear payoff distribution. They perceive a game structure like that in Fig. 6.6f when the original game has a linear or close to linear payoff distribution. The 'always play B^C' strategy continues to dominate for these agents.

Figure 6.6g, h visualise the structure of the games perceived by affiliation-motivated agents. We see that there are now two equilibrium outcomes, (B^C, B^C) and (B^D, B^D), with nAff(1) agents tending to prefer the latter for a greater range of values of $P^2(B_t = B^C)$.

The next section constructs some empirical demonstrations to show how these theoretical results manifest themselves during actual interactions between agents.

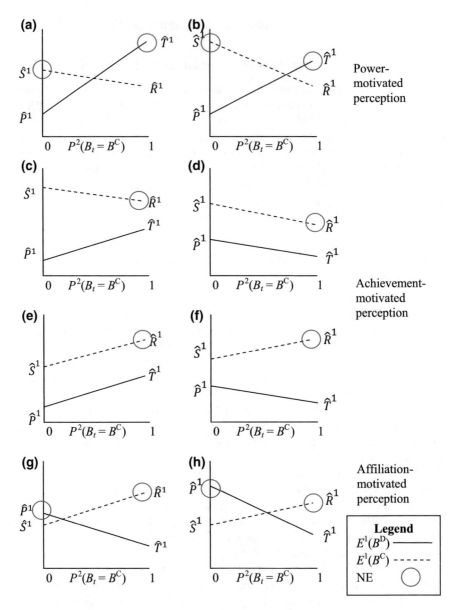

Fig. 6.6 Visualisation of the leader game perceived by **a** nPow(1) agents (Theorem A.7); **b** nPow (2) agents (Theorem A.8); **c** and **d** nAch(1) agents (Theorem A.9); **e** and **f** nAch(2) agents (Theorem A.10); **g** nAff(1) agents (Theorem A.11); **h** nAff(2) agents (Theorem A.12)

6.4.2 Empirical Study: Motivated Learning Agents as Explorers

This section examines agents with different motives playing the leader game. This section considers the same pairings of nPow(1), nAch(1) and nAff(2) agents as the empirical study in Sect. 6.3.3, but agents play a leader game rather than the PD game. That is, they are initialised in line 1 of Algorithm 3.4 with **W** defined in Eq. 6.2.

$$\mathbf{W} = \begin{bmatrix} 1 & 4 \\ 3 & 2 \end{bmatrix} \tag{6.2}$$

The context of this experiment is the mid-late stages of a TBS game when players and NPCs can see each other on the map, and must choose whether to lead exploration or focus on other activities within their own territory.

When both the NPC and the player are nPow(1) agents, they both perceive a leader game. Either the (B^C, B^D) or the (B^D, B^C) outcome is preferred over time, as shown in the top left of Fig. 6.7. The agent that leads the exploration depends on the initial probabilities with which each agent chooses B^C or B^D.

In contrast to nPow(1) agents, nAff(2) agents when opposing each other will either bilaterally explore or bilaterally pursue other activities, depending on their initial probabilities of choosing B^C and B^D (see the bottom right of Fig. 6.7). We can perhaps think of this in terms of affiliation-motivated agents wanting to exhibit the same behaviour as their peers. Wanting to belong to a group is one of the characteristics of affiliation-motivated individuals recognised by psychologists (Chap. 1, Table 1.2).

When an nPow(1) NPC encounters a player who is not similarly power-motivated, several different equilibria can emerge depending on the motive profile of the opposing player. This can be seen from column 1 of Fig. 6.7. When nPow(1) agents play nAch(1) agents, the nPow(1) agent always leads the exploration. The other player chooses an alternative activity. We see that, when opposing a nAch(1) agent that has an initial preference for leading the exploration, nPow(1) agents will initially increase their probability of executing other activities. However, once the nAch(1) agent starts to do the same, the nPow(1) agents learn to exploit this and lead the exploration. Eventually an equilibrium is reached with nPow(1) agents choosing to lead exploration (B^D) and nAch(1) agents choosing alternative activities (B^C).

We see two things from this. First we see that nPow(1) agents have a preference for winning or controlling the resources in the game. That is, by leading the exploration they are more likely to obtain collectable resources. This is one of the characteristics of power-motivated individuals recognised by psychologists (see Chap. 1, Table 1.2). Secondly, we see that nPow(1) agents will change their behaviour depending on the behaviour of their opponent. They will lead if another player has any inclination not to, but pursue alternative activities to maximise payoff if another player chooses to lead.

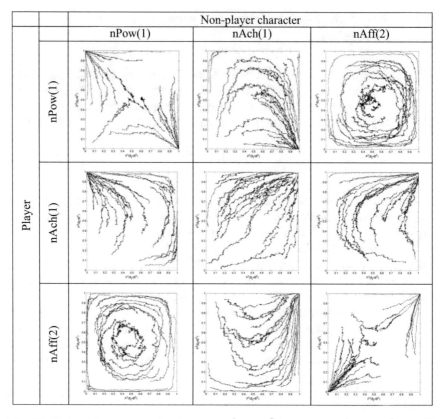

Fig. 6.7 Each subfigure shows the change in $P^1(B_t = B^C)$ (*horizontal axis*) and $P^2(B_t = B^C)$ (*vertical axis*) when 30 pairs of agents with different motives play 3,000 iterations of the leader game. See Fig. 6.4 for legend

Interestingly, a leader-follower equilibrium is not reached when nPow(1) agents encounter nAff(2) agents. As already discussed, nAff(2) agents prefer outcomes in which either both players pursue alternative activities or both players explore. Thus, when the nPow(1) agents begin to exploit nAff(2) agents by trying to lead the exploration, nAff(2) agents will learn to explore as well. An ongoing cycle emerges between the four pure strategy equilibria (see bottom left and top right of Fig. 6.7) as nAff(2) agents essentially mimic their peers.

nAch(1) agents, when playing other nAch(1) agents or nAff(2) agents, will tend to pursue other activities rather than lead exploration. In the context we have here, this means that they will choose to remain in their own territory and pursue their own civilisation growth without direct conflict with the other player. We can perhaps think of this as 'working alone', which is one of the characteristics of achievement-motivated individuals recognised by psychologists (see Chap. 1, Table 1.2).

In terms of selecting a motivated agent to control a NPC in exploration scenarios that can be modelled by a leader game, we can conclude that:

- nPow(1) agents prefer outcomes in which one leads exploration and the other follows (either (B^C, B^D) or (B^D, B^C)). Power-motivated agents prefer to be the one to lead, but can change their behaviour if there is competition for leadership.

- nAch(1) agents will tend not to lead exploration.

- nAff(2) agents will try to exhibit the same behaviour as their opponent. This may result in their changing their behaviour over time.

Adaptive diversity is again demonstrated, with different agent types exhibiting different behaviour against different opponents. The next section strives to make these empirical results more concrete by embedding the agents in a TBS simulation where the actual exploratory and activity trajectories of agents are visualised (rather than the values of the internal agent variables).

6.4.3 Application: Motivation to Explore in a Turn-Based Strategy Game

As we have seen, in TBS games, two or more opponents (some of which may be automated) take turns controlling their own empire or civilisation [14]. Control here might involve issuing commands to military units, allocating resources, or constructing new buildings. A key challenge in designing computer-controlled NPCs for TBS games is to make the NPCs' strategies appear intelligent and provide competition, while permitting the human player to win in the end [16]. Once the player has familiarized themselves with the tactics of computer-controlled NPCs, the game can rapidly become boring. Behavioural diversity between NPCs is thus desirable.

In TBS games like *Civilisation* [8] or *Call to Power II* [1], players and NPCs must control decision making concerned with hundreds of different kinds of military units, buildings, natural resources and so on. Existing work has focused on holistic planning strategies for managing this decision making [14]. The application in this section focuses only on a single decision—whether to explore a new location or improve (terraform) the current location. The intention here is to demonstrate how behavioural diversity can be achieved in decision making using motivated learning agents, with the idea that this could augment traditional planning strategies.

In this section, two agents are involved in an iterative decision-making process over the course of a TBS scenario. In any given turn, each agent has two choices: to explore a new location adjacent to its current location (B^D), or to terraform the current location (B^C). Terraforming here might mean building a road, laying services, or otherwise improving the land at the current location. The agents initially have the same probability of either behaviour, B^D or B^C. The agents use the same decision-making algorithm that we examined empirically above (Algorithm 3.4), with the same parameter settings. Both agents make 3,000 decisions. We know from the experiments in the previous section that this means that both agents are fully committed to their strategy by the end of each demonstration.

These demonstrations take place in a 50×50 tile square world as shown in Fig. 6.8. The red agent A^1 is initially located in the north and the blue agent A^2 in the south. For demonstration purposes, the red agent here is always a nAch(1) agent ($\Omega^1 = 2.6$) with $P^1(B_0 = B^C) = 0.5$. We call this a nAchNeutralAgent. We demonstrate five different opponents (blue agents) in five different experiments:

- Experiment 1: The opponent is an nPow(1) agent ($\Omega^2 = 3.9$) with $P^2(B_0 = B^C) = 0.5$. We call this an nPowNeutralAgent;

- Experiment 2: The opponent is a nAch(1) agent ($\Omega^2 = 2.6$) with an initial preference for exploring ($P^2(B_0 = B^C) = 0.25$). This agent has an initial preference for the competitive B^D behaviour, so we call it a nAchCompAgent;

- Experiment 3: The opponent is another nAchNeutralAgent as defined above.

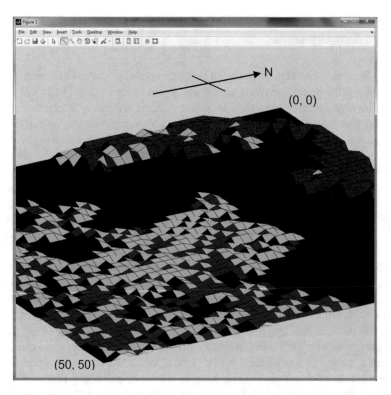

Fig. 6.8 A simulated turn-based strategy game. The *red* agent is initially located at (25, 5) and the *blue* agent is at location (25, 45). *Green* land is explored, but not terraformed. *Red* and *blue* land is terraformed by the respective agent. *Black* land is unexplored. It is assumed that the agents can see each other's activities and movements

- Experiment 4: The opponent is a nAch(1) agent with an initial preference for terraforming ($P^2(B_0 = B^C) = 0.75$). This agent has an initial preference for the cooperative B^C behaviour, so we call it a nAchCoopAgent;

- Experiment 5: The opponent is a nAff(2) agent ($\Omega^2 = 1.1$) with $P^2(B_0 = B^C) = 0.5$. We call this a nAffNeutralAgent.

The motivated agents (A^1, red, and A^2, blue) have the same decision-making processes as those examined empirically (i.e., Algorithm 3.4), but specific behaviours were implemented for this application:

- B^C terraforms the current location, changing its colour to the characteristic colour of the agent (either red or blue).

- B^D attempts to move the motivated agent to one of eight adjacent locations. The direction of movement is chosen at random, with all directions equally probable. Of course a more sophisticated exploratory strategy could be used, but this is beyond the scope of this study. The agent cannot move onto a location that has been explored or terraformed by another agent, nor can it move off the edge of the map. It can, however, move back over territory that it has explored or terraformed itself.

Thirty pairs of such agents were permitted to engage in the game, resulting in a total of 150 games. Each pair was permitted to make 3,000 decisions. We assumed that the agents were aware of the choices made by their opponents after they were made. That is, they could see each other's activities on the map. We recorded the number of locations explored (but not terraformed) and the number of locations terraformed by each agent.

Table 6.3 presents the mean and standard deviation of the number of locations explored (without being terraformed) and the number of locations terraformed during the agent interactions. We also discuss some results in terms of the number of locations 'visited', where 'visited' is defined as the sum of the number of locations terraformed and the number of locations explored but not terraformed.

We first consider the red nAchNeutralAgents. As described above, these agents use the same agent model with the same parameter settings against all opponents. However, we can see from Fig. 6.9 that their behaviour varies depending on their opponent. In particular, a two-sample z-test reveals a statistically significant difference at the 95 % confidence level in the emergent behaviour of nAchNeutralAgents when they play nPowNeutralAgents, nAchCompAgents or other nAchNeutralAgents, compared to when they are playing nAchCoopAgents or nAffNeutralAgents. Specifically the nAchNeutralAgents end the game with statistically larger areas of terraformed territory when playing nAchCoopAgents or nAffNeutralAgents than when they play nPowNeutralAgents, nAchCompAgents or other nAchNeutralAgents.

We now turn our attention to the various blue agents. On average, nPowNeutralAgents visit 26 % of locations on the map and terraform close to half of the locations they visit. In contrast, their nAchNeutralAgent opponents visited

Table 6.3 Average number of locations terraformed or explored (but not terraformed) by different types of agents, including standard deviation

Expt	Agent summary		Average (and standard deviation) of number of locations	
			Terraformed	Explored
1	Red	nAchNeutralAgent	198.2 (33.6)	30.5 (11.6)
	Blue	nPowNeutralAgent	319.7 (28.9)	331.6 (85.0)
2	Red	nAchNeutralAgent	209.7 (35.7)	38.2 (15.0)
	Blue	nAchCompAgent	307.9 (43.6)	120.7 (42.7)
3	Red	nAchNeutralAgent	217.4 (44.9)	39.3 (17.9)
	Blue	nAchNeutralAgent	216.4 (39.4)	38.3 (14.8)
4	Red	nAchNeutralAgent	232.7 (45.9)	38.7 (16.1)
	Blue	nAchCoopAgent	122.0 (33.9)	10.0 (5.9)
5	Red	nAchNeutralAgent	221.8 (37.0)	38.6 (16.7)
	Blue	nAffNeutralAgent	232.6 (43.9)	49.7 (20.0)

Agents have equal initial preference for terraforming and exploration, except where indicated

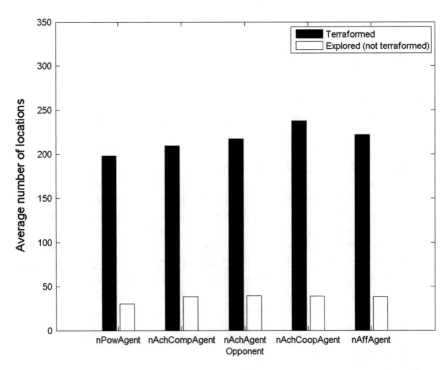

Fig. 6.9 Comparison of the behaviour of five groups of 30 nAchNeutralAgents against different types of opponents. See Table 6.3 for salient statistics

Fig. 6.10 Experiment 1:
Sample behaviour when a
nAchNeutralAgent (*red*)
opposes an
nPowNeutralAgent (*blue*)

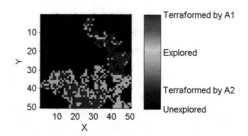

only 9 % of locations, but terraform 87 % of those locations. Figure 6.10 shows a
sample outcome when a nAchNeutralAgent (red) opposes an nPowNeutralAgent
(blue). The nPowNeutralAgent explores almost half the map in this case. In contrast, the nAchNeutralAgent explores less of the map, but has terraformed almost all
of the territory it has explored.

In general we see in Fig. 6.11 that exploration and terraforming are linked to
some extent. Even agents that demonstrate a preference for the B^C behaviour in the
empirical experiments are limited in their emergent ability to terraform land if they
are not willing to explore to find it.

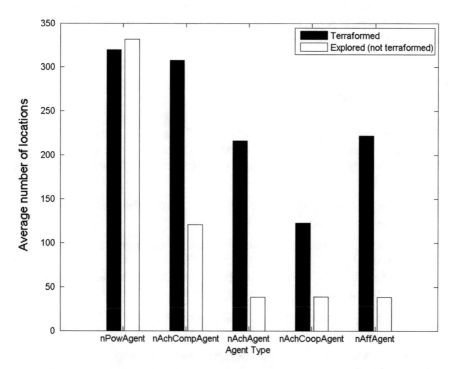

Fig. 6.11 Comparison of the behaviour of five groups of 30 agents of different types against
nAchNeutralAgents. See Table 6.1 for salient statistics

When achievement-motivated agents oppose each other, the outcome is affected by their initial preference for exploration. Specifically, nAchCompAgents will explore more compared to nAchCoopAgents. However, these initial preferences, $P^j(B_0 = B^C)$, are clearly not the sole driver of behaviour. Our nPowNeutralAgents have equal initial preference for both exploration and terraforming and still explore more on average than a nAchCompAgent with a higher initial preference for exploration. This is because the motivation of an nPowNeutralAgent to seek out competitive scenarios is stronger than that of a nAchCompAgent, regardless of how the initial probability of choosing B^C is set. This supports the empirical results in Sect. 6.4.2, which show that OMI is a more influential control parameter for emergent behaviour than $P(B_0 = B^C)$. It can change the long-term strategy adopted by the agent, where $P^j(B_0 = B^C)$ can only influence the speed with which that strategy stabilises.

Opposing nAchNeutralAgents and nAffNeturalAgents tend to focus heavily on terraforming. Statistical results show no significant difference in the average behaviour of nAchNeutralAgents and nAffNeutralAgents. This is supported by the empirical results in Sect. 6.4.2.

We see from this demonstration how motivation manifests itself in behaviour, and how a diversity of behaviour results. Not only do agents with different motives have statistically different behaviour, but one agent type can have measurably different behaviour against different opponents. The advantage of this is that it will be more difficult for players to predict the behaviour of an agent because their own play style will influence the way it responds.

6.5   Conclusion

This chapter has demonstrated the application of motivated agents in a number of abstract and concrete computer game scenarios. To conclude the study, we derive here a series of general steps that can be followed to apply motivated agents in a game scenario:

1. Select the in-game scenario or mini-game. Identify the types of NPCs in the scenario. Look for:

   a. Similarities between the scenario and abstract games described in this book;
   b. Similarities between the existing agent technologies and the extended, motivated agent algorithms in this book.

2. Using cues (a) and (b) above, identify the motivated agent algorithm to apply (refer to Chap. 3);

3. Select values for algorithm parameters (refer to Chap. 2 for theory and Part III for examples).

In general, we have seen that application of agents with different motives results in:

- Power-motivated agents that take risks or exploit their opponents. Power-motivated agents may also exhibit characteristics of Bartle's 'killer' player type.

- Achievement-motivated agents that are moderate risk takers. They may be more adaptable, changing their behaviour depending on their opponent.

- Affiliation-motivated agents that are cooperative and may tend to choose the same strategy as their opponent. Affiliation-motivated agents have the greatest misperception of a game or scenario. Their behaviour is thus the most unusual in terms of what might be considered typical game strategies.

Overall, we can see some of the characteristics that are recognised by psychologists reflected in the behaviour of our motivated agents (refer to Chap. 1, Table 1.2). Thus, we have achieved not only behavioural diversity, but some sympathy with the broad characteristics of behaviour influenced by different motives in humans. This provides a distinctly novel approach to the design of NPCs for computer games or other kinds of virtual environments. We can produce a cast of diverse characters, for each of which we can predict the strategy of play. Because the strategy of play is predictable, we have the opportunity for a higher-level reasoning process, in turn to influence the deployment of agents with different motives. We study such a scenario in Chap. 9.

Acknowledgments The paratrooper game, including the addition of motivated rule-based agents, described in Sect. 6.2 was implemented by Ben Quinton as part of a second year electrical engineering project at UNSW Canberra.

References

1. Call to Power II, *Activision* (2000)
2. R. Bartle, Hearts, clubs, diamonds, spades: Players who suit MUDs. J. Virtual. Environ. **1** (online) (1996)
3. A. Colman, *Game theory and experimental games: the study of strategic interaction* (Pergamon Press, Oxford, England, 1982)
4. M. Eisen, Lance Armstrong and the prisoners' dilemma of doping in professional sport, *Wired* (2012) http://www.wired.com/2012/10/lance-armstrong-and-the-prisoners-dilemma-of-doping-in-professional-sports/. Accessed 12 Sept 2014
5. D. Johnson, P. Stopka, J. Bell, Individual variation evades the prisoner's dilemma. BMC Evol. Biol. **2** (2002)
6. G. Kuperberg, Paratrooper. Orion Softw. (1982)
7. J. Laird, M. van Lent, Interactive computer games: Human-level AI's killer application, in *Proceedings of the National Conference on Artificial Intelligence (AAAI)*, 2000, pp. 1171–1178
8. S. Meier, Civilization, http://www.civilization.com/en/home *2K Games* (1991)
9. S. Meier, Alpha Centauri, *Electronic Arts* (1999)

10. K. Merrick, K. Shafi, A game theoretic framework for incentive-based models of intrinsic motivation in artificial systems, in *Frontiers in Psychology, Cognitive Science, Special Issue on Intrinsic Motivations and Open-Ended development in Animals, Humans and Robots,* vol 4, 2013

11. K. Merrick, The role of implicit motives in strategic decision-making: Computational models of motivated learning and the evolution of motivated agents. GAMES, Special Issue on Psychological Aspects of Strategic Choice **6**, 604–636 (2015)

12. W. Poundstone, *Prisoner's Dilemma* (Doubleday, New York, NY, 1992)

13. B. Quinton, A methodology for applying motivated agents in games., in *Electrical Engineering Second Year CDF Project, School of Engineering and Information Technology* (University of New South Wales, Canberra, 2015)

14. A. Sanchez-Ruiz, S. Lee-Urban, H. Munoz-Avila, B. Diaz-Agudo, P. Gonzalez-Calero, Game AI for a turn-based strategy game with plan adaptation and ontology-based retrieval, in *Proceedings of the ICAPS-2007 Workshop on Planning in Games,* 2007

15. J. Van Caneghem, Heroes of Might and Magic, *New World Computing* (1995)

16. E. Welch, Designing AI algorithms for turn-based strategy games, *Gamasutra* http://www.gamasutra.com/view/feature/129959/designing_ai_algorithms_for_.php. Accessed 27 July 2007

17. N. Yee, Motivations of play in online games. Cyberpsychol. Behav. **9**, 772–775 (2007)

Chapter 7
Pets and Partner Characters

Pets and partners are the friends and allies that help players overcome their adversaries. In many cases, pets and partners will follow a player unconditionally. In this chapter we explore the consequences when pets and partners have the option of satisfying their own motives. To this end, this chapter considers two scenarios for motivated learning agents as pets and partner characters. These scenarios are analysed theoretically, via an empirical investigation and through an application to battle pets. We show how motivated agents can provide us with a diversity of behaviour, and how the theoretical results permit us to predict the outcomes in an application.

7.1 Pets, Partners and Minions

Partner characters are non-player characters that can be recruited by a player to assist them play a game. Such characters are common in combat-oriented games such as *Warcraft* [1] and *World of Warcraft* [4]. *Warcraft* offers a range of 'battle pets', while *World of Warcraft* offers certain types of player characters a summonable 'minion' to assist in battle. Other kinds of pets are also offered as rewards for special achievements such as time spent playing the game. These pets may have behaviour as simple as following their owner.

Partner characters are provided as friends and allies with whom player characters can cooperate to achieve the goals of the game. Learning is a particularly desirable characteristic for partner characters because it makes possible a new dimension of cooperation. In the human-computer cooperation literature, cooperation between two entities means that both entities 'interfere with', influence and modify each other's goals, resources and behaviour for mutual benefit [6]. We argue here that an embedded learning algorithm is critical for the behaviour of a non-player character to be genuinely subject to such modification over time.

© Springer International Publishing AG 2016
K.E. Merrick, *Computational Models of Motivation for Game-Playing Agents*,
DOI 10.1007/978-3-319-33459-2_7

This chapter considers game-theoretic representations for two common inter-actions between a player and a partner character. The first scenario, in Sect. 7.2, gives the player a choice of whether to fight together with their partner, or send the partner to battle alone. This is a common decision made when playing with min-ions. We use the chicken game to abstract this interaction. Section 7.3 considers a scenario in which both a pet and its owner are able to choose a goal and must negotiate which goal to pursue. A goal here might be a quest, or a puzzle to solve. The battle of the sexes game is used to abstract this scenario. The first application demonstrates how non-player characters with different motives learn different strategies for cooperation, some more exploitative than others. The latter applica-tion demonstrates a simple design for a pet or partner character that can have its goals influenced by the actions of its owner. Both applications demonstrate a diversity of behaviour across agents with different motives.

This chapter focuses on another two two-by-two mixed motive games. Like the games studied in Chap. 6, these games again have the matrix form discussed in Chap. 3 (see Table 3.1 and Eq. 3.12), but with different payoff structures. Specifically:

- Chicken (or snowdrift): $T > R > S > P$ and

- Battle of the sexes: $S > T > R > P$.

Both games have a number of definitions in the literature, but we use the definition presented by Colman [2] here, in which none of the payoff values are permitted to be equal.

7.2   The Chicken (or Snowdrift) Game

In the well-known 'dangerous game' of chicken, two motorists speed towards each other on a collision course. Each has the option of swerving to avoid a collision, and thereby showing themselves to be 'chicken' (B^C), or of driving straight ahead (B^D). The payoff if both players are 'chicken' is $V^1 = V^2 = R$. If only one player is 'chicken' and the other drives straight on, then the 'chicken' loses face and the other player, the 'exploiter', wins a prestige victory. The payoff if A^2 drives in the hope that A^1 will be 'chicken' is $V^2 = T$. The payoff for Player 1 in this scenario is $V^1 = S$. If both players drive a collision will occur. The payoff for this outcome is the same for both players: $V^1 = V^2 = P$. The game of chicken has also been used to model real-world scenarios in national and international politics involving bilateral threats, as well as animal conflicts and Darwinian selection of evolutionarily stable strategies [2].

The rules of chicken also apply to another abstract game called snowdrift. The snowdrift game occurs when two drivers are stuck on either side of a snowdrift. Each driver has the option of shovelling snow to clear a path (B^C), or remaining in their car (B^D). The highest payoff outcome is from leaving the opponent to clear all the snow. However, if neither player clears the snow (the (B^D, B^D) outcome), then neither can traverse the drift. This is thus the lowest payoff outcome.

7.2.1 Perception of the Snowdrift Game by Motivated Learning Agents

This section draws on the snowdrift interpretation to demonstrate attribution of personalities to a non-player partner character (NPC). Suppose that a (human) player can recruit a NPC to assist with fight sequences in a game. When the pair go into battle, each has a choice: to enter the fray (B^C) or to hang back (B^D). If both players enter the fray (the (B^C, B^C) outcome), they will likely win the battle, but may be injured or their equipment damaged. The highest incentive outcome is thus to hang back and let one's partner fight the battle. However, if neither player fights, the battle cannot be won and there will be no loot to share. Usually this is a decision made by the player character—whether to send their partner to battle alone or to fight together. The introduction of computational motivation in learning agents permits the NPC to make this decision simultaneously and to change its behaviour over time in response to the actions of its player partner.

We can follow the same process as outlined in Chap. 6 and Appendix A to construct the perceived versions of snowdrift. For the snowdrift game agent types are defined by the following optimally motivating incentive (OMI) ranges:

1. nPow(1) agents: strong power-motivated agents, $T > \Omega^j > 1/2(T + R)$

2. nPow(2) agents: weak power-motivated agents, $1/2(T + R) > \Omega^j > 1/2(T + S)$

3. nAch(1) agents: achievement-motivated agents, $1/2(T + S) > \Omega^j > 1/2(S + R)$

4. nAch(2) agents: achievement-motivated agents, $1/2(S + R) > \Omega^j > 1/2(R + P)$

5. nAff(1) agents: weak affiliation-motivated agents, $1/2(R + P) > \Omega^j > 1/2(S + P)$

6. nAff(2) agents: strong affiliation-motivated agents, $1/2(S + P) > \Omega^j > P$

Figure 7.1 visualises the structure of the eight games that can be perceived when motivated agents engage in a scenario modelled by a snowdrift (or chicken) game. Figure 7.1 shows that the perceived payoff structure of the game changes as OMI decreases. nPow(2) agents, for example, perceive a game in which the 'always play B^C' strategy dominates (see Fig. 7.1b). Figure 7.1c–g show that the strategy of always playing B^C continues to dominate for all achievement-motivated agents and nAff(1) agents, regardless of the distribution of payoff in the original game.

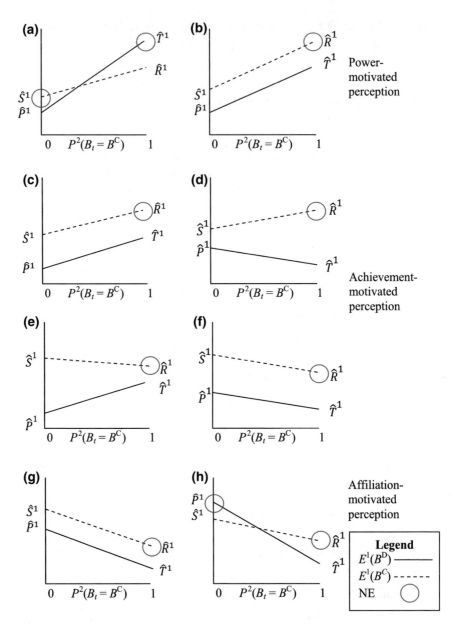

Fig. 7.1 Visualization of the snowdrift games perceived by **a** nPow(1) agents (Theorem A.13); **b** nPow(2) agents (Theorem A.14); **c** and **d** nAch(1) agents (Theorem A.15); **e** and **f** nAch(2) agents (Theorem A.16); **g** nAff(1) agents (Theorem A.17); **h** nAff(2) agents (Theorem A.18). Nash equilibria (NE) are *circled*

Figure 7.1h visualises the structure of the game perceived by nAff(2) agents. We see that there are now two equilibrium outcomes, (B^C, B^C) and (B^D, B^D). The specific outcome that emerges as the equilibrium depends on the initial probability with which the opponent chooses B^C.

7.2.2   Empirical Study: Motivated Learning Agents as Partner Characters

This section considers the same pairings of nPow(1), nAch(1) and nAff(2) agents as the empirical study in Sect. 6.3.3, but agents play a snowdrift game rather than the PD game. They are initialised in line 1 of Algorithm 3.4 with $\mathbf{W}$ defined in Eq. 7.1.

$$\mathbf{W} = \begin{bmatrix} 1 & 4 \\ 2 & 3 \end{bmatrix} \tag{7.1}$$

As in Chap. 6, the initial probabilities $P^1(B_0 = B^C)$ and $P^2(B_0 = B^C)$ required to initialise the agents in line 3 of Algorithm 3.4 are selected randomly from a uniform distribution. $P^1(B_0 = B^D) = 1 - P^1(B_t = B^C)$ and $P^2(B_0 = B^D) = 1 - P^2(B_t = B^C)$. $\alpha = 0.001$.

We again examine the behaviour of agents interacting and learning during 3,000 iterations of each game. As in Chap. 6, the results in this section are presented visually in terms of the change in probabilities of agents choosing B^C (and by implication, B^D). We again assume that one agent represents a player and the other an NPC. The horizontal axis on each chart measures $P^1(B_t = B^C)$ for the NPC and the vertical axis measures $P^2(B_t = B^C)$ for the player. Thus, a point in the upper right corner of each graph is indicative that both agents are choosing B^C. The other corners also have a prescribed meaning (refer to Fig. 6.4 in Chap. 6 for these definitions). Points closer to the centre of the chart are indicative of a more or less equal preference for B^C and B^D.

When both the player and the NPC are nPow(1) agents, they both perceive a snowdrift game. Either the (B^C, B^D) or the (B^D, B^C) outcome is learned over time as shown in the top left of Fig. 7.2. Which player enters the fray depends on the initial probabilities of choosing B^C or B^D.

When the NPC is an nPow(1) agent and the player is not, different equilibria emerge, as shown in column 1 of Fig. 7.2. When nPow(1) NPCs partner with nAch (1) players, the nPow(1) agent always learns to refuse to fight and the player must learn to fight. We see that, when partnered with a nAch(1) agent that has an initial preference for hanging back, nPow(1) NPCs will initially increase their probability of fighting. However, once the nAch(1) agent starts to do the same, the nPow(1) agents learn to exploit this and hang back. Eventually an equilibrium is reached with nPow(1) agents choosing B^D and nAch(1) agents choosing B^C.

This equilibrium is not reached when nPow(1) agents encounter nAff(2) agents. nAff(2) agents prefer outcomes in which both players do the same thing (see the bottom right of Fig. 7.2). In this case, both players cooperate and fight together or

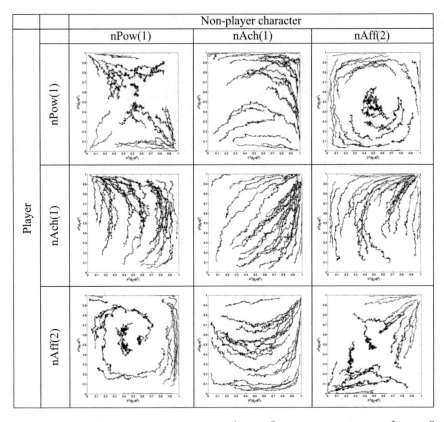

Fig. 7.2 Each subfigure shows the change in $P^1(B_t = B^C)$ (*horizontal axis*) and $P^2(B_t = B^C)$ (*vertical axis*) when 30 pairs of agents with different motives play 3,000 iterations of the chicken/snowdrift game. See Fig. 6.4 for legend

both make the decision to hang back. Thus, when the nPow(1) agents begin to exploit the nAff(2) agents, the nAff(2) agents will learn to refuse to fight. An ongoing cycle emerges between the four pure strategy equilibria (see bottom left and top right of Fig. 7.2). nAch(1) agents (middle column) are universally cooperative and will join the fight with any partner, even if their partner exploits this goodwill by hanging back.

In terms of selecting a motivated agent to control a partner character in scenarios that can be modelled by a snowdrift game, we can conclude that:

- nPow(1) agents prefer outcomes in which one partner hangs back (either (B^C, B^D) or (B^D, B^C)). If the partner is not power-motivated then the power-motivated agent will be the one that hangs back. That is, the power-motivated agent will gain the benefit from winning the fight, without injury or damage. This is consistent with the preference for competitive behaviour identified in Chap. 1 (Table 1.2).

- nAch(1) agents will join the battle, regardless of the motive profile of their partner.

- nAff(2) agents prefer the outcomes where both players do the same thing. This is consistent with wanting to belong to a [peer] group, as in Table 1.2.

Adaptive diversity is again demonstrated, with different agent types exhibiting different behaviour against different opponents. The adaptive characteristics of the three types of agents playing snowdrift are slightly different to those when they play a PD or leader game (see Chap. 6), as they are automatically tailored to the incentives offered in the current game.

7.2.3 Application: Motivated Battle Pets or Minions

The experiments in the previous section show the convergence of behavioural probabilities during iterative decision making. This section presents demonstrations in which the iterative decision-making process occurs during a single attack. That is, two agents, initially with the same probability of joining the fight, either continue to charge in or pull up short, depending on their motives and the behaviour of their partner. The agents use the same decision-making algorithm that we examined in the previous section, with the same parameter settings ($\alpha = 0.001$). Both agents make 3,000 decisions as they approach their enemy. We know from the experiments in the previous section that this means that both agents are fully committed to their strategy by the end of each demonstration. On average, in a MATLAB simulation it takes an agent just under four seconds to make 3,000 decisions and execute an action for each of those decisions, including the display of its new position on the screen.

7.2.3.1 Scenario Setup

These experiments take place in a 16×16 m, square room as shown in Fig. 7.3. Two partners (the blue agent A^1 in the south-west and the green agent A^2 in the south-east) have entered the room through the door at position $\mathbf{d} = [8, 0]$. Their initial positions are $\mathbf{p}_0^1 = [6, 2]$ and $\mathbf{p}_0^2 = [10, 2]$. Once in the room, they encounter an enemy (the red agent A^3 to the north of the room at location $\mathbf{p}_0^3 = [8, 12]$). For demonstration purposes, the blue agent here is always achievement-motivated ($\Omega^1 = 2.6$) with $P^1(B_0 = B^C) = 0.5$. This is a nAchNeutralAgent, as defined in Chap. 6. We know from the results in Sect. 7.2.2 that such an agent will always commit to an attack, regardless of the motives of its partner. We demonstrate four different partners (green agents):

Fig. 7.3 Two partners (*blue* in the southwest and *green* in the southeast) have entered through the door at the south and encountered an enemy (*red*) in the north

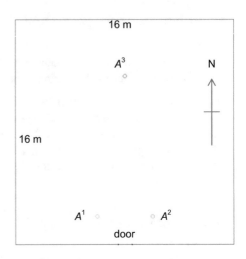

- Experiment 1: the partner is an nPow(1) agent ($\Omega^2 = 3.9$) with $P^2(B_0 = B^C) = 0.5$. We call this an nPowNeutralAgent;
- Experiment 2: The partner is a nAch(1) agent ($\Omega^2 = 2.6$) with an initial preference for hanging back $P^2(B_0 = B^C) = 0.25$. This agent has an initial preferences for exploiting its partner, so we call it a nAchCompAgent;

- Experiment 3: The partner is a nAch(1) agent ($\Omega^2 = 2.6$) with an initial preference for a cooperative attack $P^2(B_0 = B^C) = 0.75$. This agent has an initial preference for cooperation so we call it a nAchCoopAgent;

- Experiment 4: The partner is a nAff(2) agent ($\Omega^2 = 1.1$) with $P^2(B_0 = B^C) = 0.5$. We all this a nAffNeutralAgent.

The motivated agents (A^1, blue, and A^2, green) in this section have the same decision-making processes as those examined in Sect. 7.2.2, but specific behaviours were implemented for this application. B^C moves the motivated agent towards the enemy location according to:

$$\mathbf{p}^j_{(t+1)} = \mathbf{p}^j_t + \lambda P_t(B_t = B^C)(\mathbf{p}^3_t - \mathbf{p}^j_t) + \phi, \tag{7.2}$$

where λ is the step size between 0 and 1 (we used $\lambda = 0.05$) and ϕ is a small number selected from a uniform distribution on the interval $[-0.05, 0.05]$. This movement update moves the agent approximately in the direction of the enemy by a distance that is proportional to the chosen step size λ and the agent's preference for such movement, represented by $P_t(B_t = B^C)$.

B^D moves the motivated agent towards the door according to:

$$\mathbf{p}^j_{(t+1)} = \mathbf{p}^j_t + \lambda P_t(B_t = B^C)(\mathbf{d} - \mathbf{p}^j_t) + \phi, \tag{7.3}$$

where **d** is the position of the door. For the purpose of this demonstration, the red enemy agent does not move. It stands firm and faces its attacker(s). Formally, $p^3_{t+1} = p^3_0$ for all t.

7.2.3.2   Discussion

Figures 7.4, 7.5 and 7.6 show samples of the resulting behaviour of some examples of the different agents. The coloured circles show their movement trajectories. The black circles show the final positions of the two partner agents. We see that the blue nAchNeutralAgent enters the fight with all four partners, although its movement trajectory is slightly different with different partners. Table 7.1 shows that on average, after 3,000 decisions, the nAchNeutralAgent (blue) ends up within 1.3 m of the enemy.

Figure 7.4 shows the movement trajectories of a nAchNeutralAgent (blue) partnered with an nPowNeutralAgent (green). The nAchNeutralAgent joins the fight after some initial hesitation. The nPowNeutralAgent approaches the fight, but then backs off towards the door, leaving its achievement-motivated partner to complete the fight alone. On average the nPowNeutralAgent will end up 9.1 m away from the enemy after the four-second decision-making sequence.

When achievement motivated agents are partnered with each other, as in Fig. 7.5, both agents will eventually join the fight. However, the trajectory they follow to do this is affected by the agents' initial preference for cooperation. In Fig. 7.1a, b the blue agents are nAchNeutralAgents. However, in Fig. 7.1a the green agent is a nAchCompAgent, while in Fig. 7.1b the green agent is a nAchCoopAgent. We see in that the nAchCoopAgent joins the fight more quickly, while the nAchCompAgent will spend more time retreating towards the door before eventually joining the fight. nAchCoopAgents will end up closer to the enemy on

Fig. 7.4 Experiment 1: Sample behaviour when a nAchNeutralAgent (*blue*) partners with an nPowNeutralAgent (*green*)

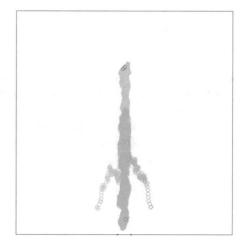

Fig. 7.5 Sample behaviour when a nAchNeutralAgent (*blue*) partners with other achievement-motivated (*green*) agents with different initial preferences for cooperation. In **a** Experiment 2: the *green* agent is a nAchCompAgent, while in **b** Experiment 3: the *green* agent is a nAchCoopAgent

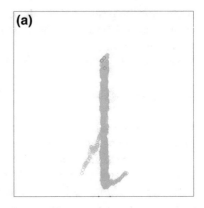

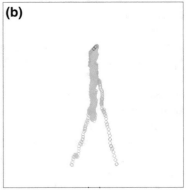

average (0.4 m). nAchCompAgents will end up further from the enemy on average (4.2 m), although not to the same extent or at the same level of reliability as a nPowNeutralAgent (note the higher standard deviation).

When a nAchNeutralAgent is partnered with a nAffNeutralAgent, as shown in Fig. 7.6, both agents will enter the fight. The sample demonstration in Fig. 7.6 shows the nAffNeutralAgent will be more reluctant to do so. However, it will end up at a similar distance from the enemy (see Table 7.1).

We see from this demonstration how motivation manifests itself in behaviour, and how a diversity of behaviour results. While other work has considered personality traits in target tracking settings [3], the methodology proposed in this book is unique in its basis on a psychological theory of motivation. Behavioural diversity is achieved by setting a single parameter, the OMI. The resulting behaviour of agents can be predicted in advance from a game-theoretic analysis of the models for the chosen OMI, but emerges over time as a result of learning.

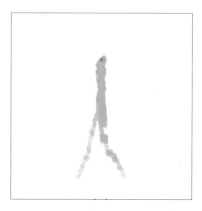

Fig. 7.6 Experiment 4: Sample behaviour when a nAchNeutralAgent (*blue*) partners with a nAffNeutralAgent (*green*)

Table 7.1 Average distance to enemy after 3,000 decisions

Expt	Agent summary		Average distance to enemy and (Standard deviation)
1	Blue	nAchNeutralAgent	1.3 (1.7)
	Green	nPowNeutralAgent	9.1 (1.7)
2	Blue	nAchNeutralAgent	1.5 (2.2)
	Green	nAchCoopAgent	0.4 (0.8)
3	Blue	nAchNeutralAgent	1.2 (1.7)
	Green	nAchCompAgent	4.2 (3.6)
4	Blue	nAchNeutralAgent	1.2 (1.6)
	Green	nAffNeutralAgent	3.2 (2.0)

Statistics computed from 30 runs of each experiment

7.3   Battle of the Sexes

The battle of the sexes game can be thought of as modelling a predicament between two friends with different interests in entertainment. Each prefers a certain form of entertainment that is different from the other's, but both would rather go out together than alone. If both opt for their preferred entertainment, leading to a (B^C, B^C) outcome, then each ends up going alone. The payoff for this outcome is $V^1 = V^2 = R$. A worse outcome (B^D, B^D) results if both make the sacrifice of going to the entertainment they dislike as they both end up alone. The payoff for this outcome is $V^1 = V^2 = P$. If, however, one chooses their preferred entertainment and the other plays the role of 'hero' and makes the sacrifice of attending the entertainment they dislike, then the outcome is better for both of them ($V^1 = T$ and $V^2 = S$ or $V^1 = S$ and $V^2 = T$). The payoff matrix for the battle of the sexes game ($S > T > R > P$) is relatively similar to that of the leader game, with the only difference in the definition being the relationship between T and S. In the leader game, $T > S$, while in the battle of the sexes game, $S > T$. This reflects the real-world relationship that is often perceived between leadership and sacrifice [7].

7.3.1 Perception of Battle of the Sexes by Motivated Learning Agents

This section uses battle of the sexes to model a negotiation between a player character and a pet. In this scenario we assume both the pet and its owner are capable of selecting a goal. The pet and its owner must then negotiate whether they should each follow their own goals, or whether one will (at least temporarily) abandon its goal to pursue the other's goal. If both the owner and the pet follow their own goals, they will have to part ways. Likewise, if each pursues the goal of the other, they will also part ways. If they wish to remain together, exactly one of the two must agree to pursue the goal of the other.

We use the following mapping of OMIs to motivation type:

1. nPow(1) agents: strong power-motivated agents, $S > \Omega^j > 1/2(S+T)$

2. nPow(2) agents: weak power-motivated agents, $1/2(S+T) > \; > \Omega^j > 1/2 (S+R)$

3. nAch(1) agents: achievement-motivated agents, $1/2(S+R) > \Omega^j > 1/2 (T+R)$

4. nAch(2) agents: achievement-motivated agents, $1/2(T+R) > \Omega^j > 1/2 (T+P)$

5. nAff(1) agents: weak affiliation-motivated agents, $1/2(T+P) > \Omega^j > 1/2 (R+P)$

6. nAff(2) agents: strong affiliation-motivated agents, $1/2(R+P) > \Omega^j > P$

Figure 7.7 visualises the structure of the eight games that can be perceived when motivated agents engage in a scenario modelled by a battle of the sexes game. We see that nPow(2) agents misperceive the battle of the sexes (Fig. 7.7b) as a leader game. Thus, although there are still two equilibria, (B^C, B^D) and (B^D, B^C), Player 1 prefers the former of these when their OMI is lower. That is, they prefer to be the player following their own goal, rather than sacrificing their goal to assist another player.

Figure 7.7c, d visualises the structure of the games perceived by nAch(1) agents. The game $\hat{T}^j > \hat{R}^j > \hat{S}^j > \hat{P}^j$ occurs for a linear or close to linear payoff distribution, while $\hat{T}^j > \hat{R}^j > \hat{P}^j > \hat{S}^j$ occurs with a highly nonlinear payoff distribution. We see

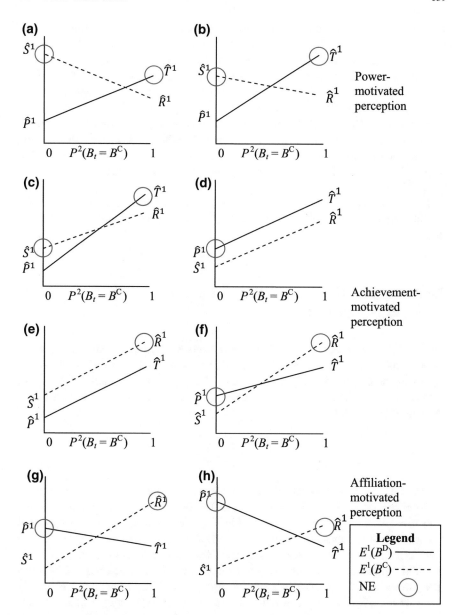

Fig. 7.7 Visualization of the battle of the sexes games perceived by **a** nPow(1) agents (Theorem A.19); **b** nPow(2) agents (Theorem A.20); **c** and **d** nAch(1) agents (Theorem A.21); **e** and **f** nAch(2) agents (Theorem A.22); **g** nAff(1) agents (Theorem A.23); **h** nAff(2) agents (Theorem A.24). Nash equilibria (NE) are *circled*

that nAch(1) agents playing a battle of the sexes game with a highly nonlinear payoff structure actually perceive a prisoners' dilemma game, rather than a battle of the sexes game. The strategy of always playing B^D dominates for these agents—noting of course that B^D here is a strategy of sacrificing one's goals, rather than of exploiting one's partner. When the payoff structure is closer to linear, a game with two equilibria, (B^D, B^C) and (B^C, B^D), is perceived.

Figure 7.7e, f visualises the structure of the perceived games for nAch(2) agents. We see that when these agents play a game with a highly nonlinear payoff structure, they perceive a game where the strategy of always playing B^C dominates. When the payoff structure is closer to linear, a game with two equilibria, (B^C, B^C) and (B^D, B^D), is perceived.

Finally, in Fig. 7.7g, h we see that for affiliation-motivated agents there are two equilibrium outcomes, (B^C, B^C) and (B^D, B^D). nAff(2) agents prefer the latter of these.

7.3.2 Empirical Study: Motivated Learning Agents as Negotiators

This section examines agents with different motives playing the battle of the sexes game. This section considers the same pairings of nPow(1), nAch(1) and nAff(2) agents as the empirical study in Sect. 6.3.1, but agents play a battle of the sexes game rather than the PD game. They are initialised in line 1 of Algorithm 3.4 with **W** defined in Eq. 7.4.

$$\mathbf{W} = \begin{bmatrix} 1 & 3 \\ 4 & 2 \end{bmatrix} \tag{7.4}$$

When both the NPC and the player are nPow(1) agents, they both perceive the battle of the sexes game. Either the (B^D, B^C) or the (B^C, B^D) outcome is preferred over time, as shown in the top left of Fig. 7.8. However, the learning trajectories towards these outcomes are not direct. Both the NPC and the player desire to follow their own goals, so their learning trajectories curve out towards the (B^C, B^C) equilibrium, before converging on the (B^D, B^C) or (B^C, B^D) equilibria. The agent that pursues its own goal depends on the initial probabilities of choosing B^C or B^D.

However, Fig. 7.8 (top row) shows that when nPow(1) agents play nAch(1) agents, a different equilibrium occurs. nAch(1) NPCs sacrifice their own goal to follow the goal of their nPow(1) partner. The vertical learning trajectories occur because nAch(1) agents learn very slowly when playing the battle of the sexes game. This occurs because the gradients of $E^2(B^D)$ and $E^2(B^C)$ are very similar (see Fig. 7.7c).

When nPow(1) and nAff(2) agents interact, a cycle emerges, because nAff(2) agents prefer the outcomes where they employ the same strategy as their opponent,

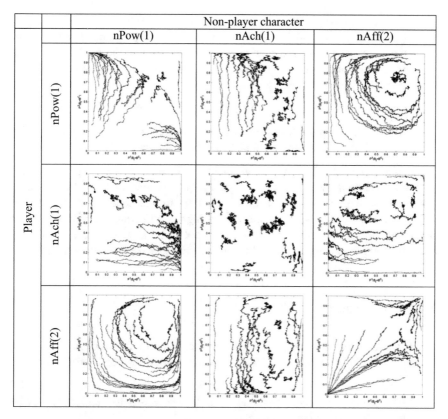

Fig. 7.8 Each subfigure shows the change in $P^1(B_t = B^C)$ (horizontal axis) and $P^2(B_t = B^C)$ (vertical axis) when 30 pairs of agents with different motives play 3,000 iterations of the battle of the sexes game. See Fig. 6.4 for legend

while nPow(1) agents do not. The same cycle emerges when nAff(2) and nAch(1) agents interact, although it is less obvious in this simulation because nAch(1) agents learn more slowly than nAff(2) agents.

When achievement-motivated agents negotiate with each other, the slow rate of learning is quite apparent, with the learning trajectories failing to reach one of the equilibria (corners). When the experiment is run for longer (10,000 time steps), these agents converge on the (B^D, B^C) or (B^C, B^D) equilibria.

In the bottom right corner of Fig. 7.8 we see that nAff(2) agents converge rapidly on the (B^C, B^C) or (B^D, B^D) equilibria, with a preference for the (B^D, B^D) equilibrium as predicted in Sect. 7.3.1. Affiliation-motivated agents like to select the same strategy as their peers, but perversely, in the battle of the sexes game, this results in the agents pursuing different goals as a result.

In terms of selecting a motivated agent to control a NPC that has the capacity to negotiate, modelled by a battle of the sexes game, we can conclude that:

- nPow(1) agents prefer outcomes in which players employ different strategies (either (B^C, B^D) or (B^D, B^C)). In the battle of the sexes game, this means that they will remain together, as one will always follow the goal of the other. When nPow(1) agents negotiate with a nAch(1) agent, the nPow(1) agent prefers to follow its own goal.

- nAch(1) agents learn very slowly playing this game.

- nAff(2) agents prefer outcomes in which players exhibit the same strategy. Perversely, this means they end up alone in the battle of the sexes game.

Adaptive diversity is again demonstrated, with different agent types exhibiting different behaviour against different opponents. We note that the behaviour of power-and affiliation-motivated agents is contrary to that predicted by psychological motivation theory. Typically, individuals high in affiliation motivation prefer to engage in activities together [5]. We conclude that the abstract nature of this scenario prevents the emergence of all of the subtleties of psychological motivation.

7.4 Conclusion

When we consider the experiments in this chapter in conjunction with those employing learning agents in Chap. 6, we see a number of trends emerging:

- Power-motivated agents will always attempt to exploit their opponents, and will generally succeed, except against other power-motivated agents. Power-motivated agents are risk takers.

- Achievement-motivated agents tend to favour cooperative solutions and more moderate risks.

- Affiliation-motivated agents will tend to mimic the behaviour of their opponent. This may result in oscillation between different strategies.

The experiments have shown that behavioural and adaptive diversity are possible using motivated learning agents. However, it is also possible to predict the long-term outcomes of learning using game-theoretic analysis. By selecting values of $P^1(B_0 = B^C)$ and $P^2(B_0 = B^C)$, game designers can control the long-term behaviour of a character, without losing the diversity that is introduced when NPCs are allowed to learn.

The applications in this chapter and Chap. 6 have shown that we do not need to change the scenario in which NPCs are embedded, nor the behaviours available to them, in order to achieve a diversity of agents. Rather, we introduce diversity into the way agents perceive a scenario. This results in diverse behavioural strategies for completing the scenario.

References

1. Warcraft, *Blizzard Entertainment*, 1994
2. A. Colman, *Game Theory and Experimental Games: The Study of Strategic Interaction* (Pergamon Press, Oxford, England, 1982)
3. S.N. Givigi, H.M. Schwartz, Swarm robot systems based on the evolution of personality traits. Turk. J. Electr. Eng. **15**, 257–282 (2007)
4. T. Graepel, R. Herbrich, J. Gold, Learning to fight, in *Proceedings of the International Conference on Computer Games: Artificial Intelligence, Design and Education*, 2004
5. J. Heckhausen, H. Heckhausen, *Motivation and Action* (Cambridge University Press, New York, NY, 2010)
6. J.-M. Hoc, From human-machine interaction to human-machine cooperation. Ergonomics **43**, 833–843 (2010)
7. B. van Knippenberg, D. van Knippenberg, Leader self-sacrifice and leadership effectiveness: the moderating role of leader prototypicality. J. Appl. Psychol. **90**, 25–37 (2005)

Chapter 8
Support Characters

Support characters are the innkeepers, drunks, blacksmiths, tailors, alchemists, merchants, soldiers and craftspeople who support the plot of a game. Support characters are often numerous, or need to interact with numerous player characters. In the first instance, crowd algorithms are particularly applicable. The first half of this chapter describes the use of a motivated crowd algorithm to control the protagonists in an original game, *Breadcrumbs*. We describe how this game abstracts other common scenarios for support characters. The second part of this chapter returns to two of the abstract games from Chap. 7 to demonstrate the use of motivated learning agents in vendors and quest givers, who need to interact with multiple player characters over time.

8.1 Motivation in Multi-agent Systems

Unlike enemies and partners, support characters do not necessarily 'takes sides' in the competitive aspects of a game. Rather, they support the plot of a game, as well as add interest and diversity. They buy and sell goods, offer services, issue quests and pass on gossip. The more numerous and diverse the support characters in a game, the richer the game world. Support characters are thus a compelling target for novel artificial intelligence algorithms, although they are among the least studied in this respect [6].

The need for large numbers of support characters makes it critical that the algorithms used to control them are computationally inexpensive. Achieving this goal, while also supporting interesting and believable behaviours, remains a challenge. This chapter explores the use of motivation to achieve diversity in crowds of motivated agents and in motivated learning agents.

As we saw in Chap. 3, crowd algorithms can produce believable behaviour via application of a set of simple rules. In this chapter we demonstrate a crowd of 250 motivated agents as the protagonists in an original game. We then explain how this

© Springer International Publishing AG 2016
K.E. Merrick, *Computational Models of Motivation for Game-Playing Agents*,
DOI 10.1007/978-3-319-33459-2_8

game can be considered an abstraction of a number of common support character scenarios that occur in large-scale game worlds. This is the topic of Sect. 8.2.

In the latter parts of this chapter we consider motivated learning agents as a computationally lightweight approach to behavioural diversity. We use the theoretical results from Chaps. 6 and 7, but discuss the outcomes when support characters interact with large numbers of different player characters over time, rather than with a single opponent. To do this, in Sect. 8.3 we introduce the idea of a player-base motive index that reflects the preferences of a group of players for conflict or cooperation.

8.2   Breadcrumbs

This section describes an original game, called *Breadcrumbs*, that uses motivated crowds. As a stand-alone game, *Breadcrumbs* is a simple, yet captivating Android App for children. However, the underlying game concept can be understood as an abstract game that can represent more complex scenarios or mini-games that might occur in a large-scale game.

8.2.1   Attracting a (Motivated) Crowd

The game is set in two rooms connected by an open doorway. The layout here is not important—we could have several rooms, or even a maze. Initially the characters (simple square-shaped *boids* in this case) are randomly distributed throughout both rooms. The rules of the game are as follows:

Aim of the game:

Place up to five breadcrumbs to lure all the *boids* into one room.

Instructions:

1. Place breadcrumbs by touching the screen at the desired location.

2. Once you have placed five breadcrumbs, you can continue placing breadcrumbs, but each new breadcrumb will trigger the removal of the oldest existing breadcrumb.

3. Breadcrumbs are always tasty—but you don't know exactly how tasty any given crumb will be. In addition, different *boids* have different preferences for flavour.

When the game begins, the *boids* begin to form small groups almost immediately, governed by Algorithm 3.3. The specific settings of the algorithm used are:

- Line 1: $n = 250$, and each *boid* has an optimally motivating incentive (OMI) between 0 and 1, generated at random from a uniform distribution. *Boids* are coloured according to their motivation type. Power-motivated agents (with OMIs between 0.66 and 1.00) are coloured red; achievement-motivated agents (with OMIs between 0.33 and 0.66) are coloured orange; affiliation-motivated agents (with OMIs between 0.00 and 0.33) are coloured yellow.

- The set of condition-goal-behaviour tuples is initially empty.

We used a flocking implementation based on software by Heavner [5] as the basis for implementing Algorithm 3.3. Heavner's implementation uses the standard *boid* rules discussed in Chap. 3, with a number of additional forces to smooth collision avoidance. A list of salient variables and their values associated with this implementation is provided in Appendix B.

Breadcrumbs in our game are attractors. When the player places breadcrumbs, a new tuple $\langle C, G, B \rangle$ is created. G is a goal, described by the position of the breadcrumb. The incentive of a breadcrumb goal $I^s(G)$ is generated at random from a uniform distribution between 0 and 1 when the breadcrumb is placed. B is assumed to be a behaviour that moves the agent towards the breadcrumb and C is a condition about when the breadcrumb will be attractive. As described in Eq. 3.9, in our game C is true if the *boid* is within a radius $R_m = 150$ pixels of the breadcrumb and the incentive of the breadcrumb is within $\gamma = 0.22$ of the agent's OMI. The implication of this definition of C is that agents may be attracted to more than one breadcrumb depending on its position, incentive and their own motivation.

Once the player begins to add breadcrumbs, the behaviour of the crowd is influenced by both the *boid* rules and any goals G^g for which C^g holds.

8.2.2 Case Studies

Breadcrumbs is, perhaps, deceptively simple. However, if we attempt an implementation without motivated agents the game is no longer compelling. If we use *boids* without motivation in such a scenario, and assume that all *boids* are the same, then we have two alternative extremes:

- All *boids* have the same deterministic goal-selection rule, for example, a greedy preference for the highest incentive. Thus all *boids* pick the same target from a given set of goals.

- All *boids* have the same probabilistic goal-selection rule. Thus, the precise goal that will be selected at a given time by a given *boid* cannot be predicted.

In the first case, behaviour is very predictable. In the second case, behaviour is unpredictable. Neither case models a systematic preference for different incentives by different agents. The use of motivation permits us to model this kind of behaviour.

The case studies below compare crowds using the two unmotivated alternatives above to motivated crowds in the *Breadcrumbs* game to explain this further.

8.2.2.1   Motivated Crowds

A number of screenshots from the version of the game using a motivated crowd are shown in Fig. 8.1. Figure 8.1a shows the agents a few seconds into the game. Small groups have already started to form. Figure 8.1b shows the state of the game after three breadcrumbs have been placed: two in the doorway, and one in open space in the bottom room. The agents have started to crowd towards the doorway and form a line between the breadcrumbs. This occurs as some of the agents initially motivated to approach the first placed crumb (towards the middle of the upper room) are then more strongly motivated to approach the second placed crumb (the one closest to the doorway). Cliques of the same (or similar) coloured agents form as agents with similar motivational preferences are attracted to the same crumbs. These cliques are particularly visible in Fig. 8.1c, which also shows a clique dispersing as its breadcrumb disappears. Occasionally, breadcrumbs will be entirely ignored.

Figure 8.1d shows the end state of the game, with all agents in one room. The game in its current form takes 5–10 min to complete. However, more complex room plans or mazes could be designed as additional 'levels' to extend the game.

One of the main challenges for the player is to keep track of which agents will disperse when a new breadcrumb is placed and be ready to manage the resulting dispersal (by placing more breadcrumbs) so as not to lose agents back through the door. The game is compelling because of the different shapes and patterns that emerge in the crowds, particularly the coloured cliques, which can be learned and utilised by the player to move the agents around the rooms.

8.2.2.2   Greedy Crowds

In this version of the game, shown in Fig. 8.2a, all agents are programmed to be attracted to any breadcrumb goal with incentive between 0.75 and 1.00. That is, the crowd is now homogeneous, and agents are greedy with respect to $I^s(G)$.

The emergent behaviour in this case still exhibits aspects of crowd behaviour, but is no longer compelling as a game. All agents crowd to the highest incentive breadcrumb(s). If lower value breadcrumbs are placed they are ignored. The player is forced to continue placing breadcrumbs until the high value breadcrumb(s) disappear. The entire crowd will then disperse and proceed to the crumb of the next highest value, if a suitable one exists. The player has no feedback on why agents ignore some crumbs and not others. In addition, the diversity of shapes and patterns observed among the motivated agents is not observed among the greedy agents.

8.2.2.3   Random Crowds

In this version of the game, all agents are programmed to be attracted to a different, randomly chosen range of breadcrumb incentives at each time step. The emergent

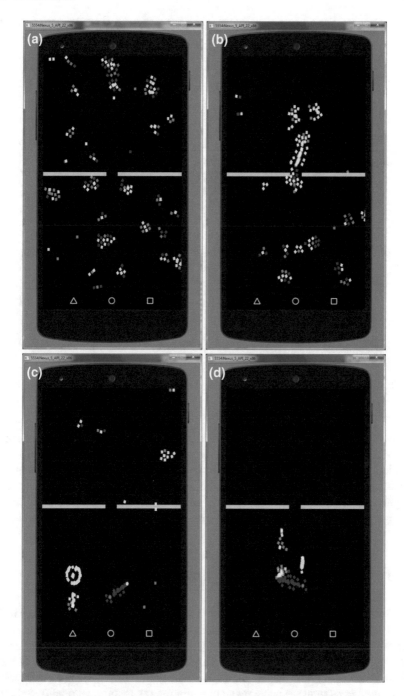

Fig. 8.1 *Breadcrumbs* **a** The initial appearance of the game: agents form small groups in both rooms; **b** a player has placed some breadcrumbs and some agents are motivated to move towards them; **c** a breadcrumb has disappeared and the agents around it are dispersing; **d** game over—all agents in one room

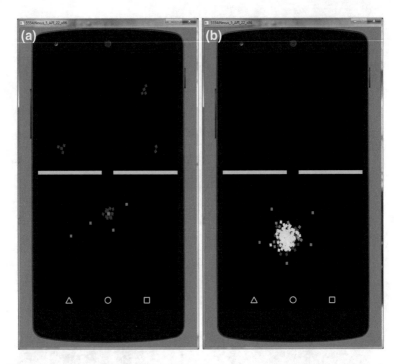

Fig. 8.2 Two variants of *Breadcrumbs* without motivated agents: **a** Greedy agents crowd around the highest incentive crumb only; **b** random agents are attracted equally to all crumbs, so their average position ends up between crumbs

behaviour exhibits some aspects of flocking, but the goal-directed aspect of crowd behaviour is lost. In this case the agents crowd aggressively towards groups of breadcrumbs, but become suspended between them. The phenomenon is shown in the screenshot in Fig. 8.2b. It occurs because the average pull on the crowd is roughly equal in the direction of each crumb. The crowd centres itself on the average position of all the crumbs as a result and the game loses its meaning.

8.2.3   Scenarios Abstracted by Breadcrumbs

The *Breadcrumbs* scenario in fact abstracts a number of situations that we see in the real world, which we might wish to build in a virtual world. For example, when games are placed on the casino floor, they are all entertaining, but different games will be preferred by different guests. The casino manager cannot tell in advance which guests will prefer which games, because their motives are not observable.

As another example, when a range of similar items of different quality and price is on display in a shop, different items will be purchased by different customers, even if all have the same function.

Finally, consider a group of performers in a circus. Each performer has a different expertise. There may be jugglers, sideshow operators, animal trainers and so on. A range of goals is thus presented to the visiting crowd, and members of the crowd will distribute themselves among the targets according to their individual motives.

In each of the examples described above, a crowd of motivated agents can be used to model the agents moving among the attractions, whether these attractions be games in a casino, merchandise in a shop or performers in a circus. The *Breadcrumbs* demonstration indicates that the following properties of the crowd will then occur:

- Formation of cliques attracted to shared goals;

- Movement of members of cliques between goals when new goals become available;

- Different cliques focusing attention on different goals.

These properties are not achieved using homogenous groups of agents, such as greedy or random crowds. With this result in mind, we now move on to a complementary question: how do motivated agents react in response to multiple players? The focus of the next section is on motivated learning agents.

8.3 Chicken Revisited

This section revisits the chicken (or snowdrift) game introduced in Chap. 7 in a support character scenario. The previous chapters included studies of equilibrium outcomes when motivated learning agents play against individual opponents. While these outcomes are interesting and useful, for many applications it is also relevant to study the long-term outcomes that occur when a motivated learning agent interacts with multiple players over the course of its life. Thus, the aim in this section is to simulate scenarios in which a single NPC interacts with multiple player characters with different motives over a long period of time. We examine vendor NPCs in this section.

8.3.1 Player-Base Motive Indices

To perform the analysis in this section, we adapt the concept of a national motive index (NMI) discussed in Chap. 1 to the gaming setting. For computer games, instead of a NMI, we assume the existence of a player-base motive index that can favour either power or affiliation motivation. We assume that if the player-base motive index favours power motivation, the general tendencies of players will be to exhibit more conflict-oriented behaviour, characteristic of power-motivated individuals. Conversely, we assume that if the player-based motivation index favours affiliation motivation then the tendencies will be to exhibit more cooperative behaviour, characteristic of affiliation motivation. We use these assumptions to focus our analysis in the next sections.

8.3.2   Motivated Learning Agents as Vendors

Vendors are an example of a support character common in many games. In games such as *World of Warcraft* [4] and *EverQuest* [3], vendors sell things such as armour, weapons, spells, food and raw materials. They also purchase unwanted items from players so they don't have to carry them around. Typically vendors buy and sell items for fixed prices that may have little variation between vendors. Vendors are often used as sinks to prevent oversupply of items, although in some games vendors do maintain an inventory of actual items purchased from players, if only for short periods.

Suppose a player wishes to negotiate a purchase from (or sale to) a vendor. The buyer and the seller must negotiate a price. The buyer would like to pay a low price, while the seller would like a high price. If both players refuse to bargain (B^D), then there will be no sale. This is the lowest payoff outcome for both, $V^1 = V^2 = P$. If one player decides to change their price (up for the buyer or down for the seller), then a sale will result. The payoff T will be higher for the one who held firm (B^D), and lower (S) for the one who wavered (B^C). If both players agree to bargain (the seller by lowering their price and the buyer by raising their offer), then a sale will occur, although not necessarily at the preferred price of either player. The payoff for both players will be R. This interaction between a player character and a non-player vendor is thus abstracted by a chicken game.

Figure 8.3 shows a dialog that permits a player to engage in a negotiation with a vendor NPC. The traditional 'buy' or 'don't buy' choices are available (the 'I'll buy it!' and 'Not today thanks' buttons respectively). A third button and associated text field permits the player to propose an alternative price. This starts a mini-game that

Fig. 8.3 Vendor dialog box for a sale. The player can either make the traditional 'buy it' or 'leave it' choices, or can start a mini-game negotiation by proposing an alternative price

conforms to the format of a chicken game. If the player clicks the 'How about…' button without increasing the value in the text box, this is a B^D choice. If the player increases the value in the text box before clicking the 'How about…' button, this is the B^C choice. The 'Not today thanks' button ends the game without a purchase being made (as does the close button at the top right of the dialog). When the player clicks the 'How about…' button, the NPC is notified of the button click, but is not immediately told whether or not a price change has been proposed (i.e. the NPC does not have any advantage of knowing whether the player has chosen B^C or B^D). The NPC then generates a B^C or B^D response based on its own $P(B_t = B^C)$. Definitions of these behaviours are:

- B^C reduces the price p_t^j of a product in the direction of the buyer's desired price. For example, the change in asking price may be proportional to a step size λ and the vendor's preference $P(B_t = B^C)$ for bargaining:

$$p_{t+1}^j = p_t^j + \lambda P\left(B_t = B^C\right)\left(p_t^3 - p_t^j\right) \tag{8.1}$$

- B^D leaves the vendor's price unchanged.

$$p_{t+1}^j = p_t^j \tag{8.2}$$

Finally, the NPC is notified of the player's decision. The possible outcomes are as follows (the NPC choice is listed first in each case):

- A (B^D, B^D) outcome leaves the dialog box unchanged.
- A (B^C, B^C) outcome updates the dialog box so that the player can see the NPC's reduced price. The NPC is notified of the player's offer.
- A (B^C, B^D) outcome updates the dialog box so that the player can see the NPC's reduced price.
- A (B^D, B^C) outcome leaves the dialog box unchanged, but the NPC is notified of the player's offer.

The mini-game ends when the player either chooses to buy the item at the (possibly reduced) price proposed by the NPC, or the player exits the dialog.

We can draw on the analysis in Chap. 7, and our assumption of a player-base motivation index, to consider the possible emergent behaviours of vendors with different motives over time.

8.3.3 Empirical Study: Vendors Bargaining with Multiple Opponents

We can use the theory discussed in Chap. 3 to predict the long-term learning outcomes of the motivated learning agents in the scenario described above. For the

purpose of the examination in the next section, we make our standard assumptions
that $T = 4$, $R = 3$, $S = 2$, $P = 1$.

We examine vendor agents with three motive profiles:

- nPow(1): a strong power-motivated agent: $\Omega^j = 3.9$

- nAch(1) : an achievement-motivated agent: $\Omega^j = 2.6$

- nAff(2): a strong affiliation-motivated agent: $\Omega^j = 1.1$

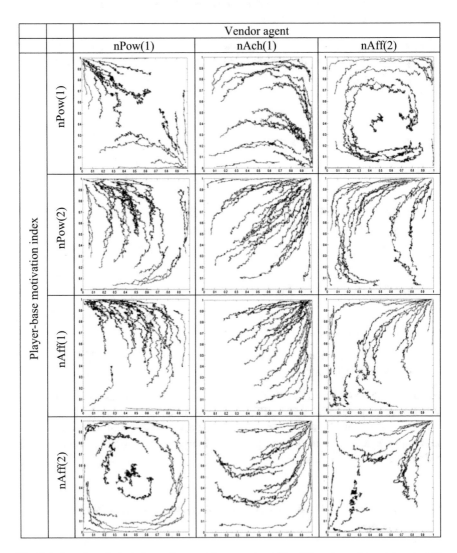

Fig. 8.4 Learning trajectories when vendor agents with different motives interact with
player-bases with different motive indices. The horizontal axis shows the change in $P(B_t = B^C)$
for the vendor agent. The vertical axis shows the change in $P(B_t = B^C)$ for the player-base. See
Chap. 6 for a description of the chart legend

We assume that the behaviour of the player-base conforms to one of the following four characteristics:

- nPow(1): a strong power-motivated player-base: $\Omega^j = 3.9$

- nPow(2) : a weak power-motivated player-base: $\Omega^j = 3.1$

- nAff(1) : a weak affiliation-motivated player-base: $\Omega^j = 1.9$

- nAff(2): a strong affiliation-motivated player-base: $\Omega^j = 1.1$

Figure 8.4 shows the learning trajectories when vendor agents with different motives interact with player-bases with different motive indices. The horizontal axis shows the change in $P(B_t = B^C)$ for the vendor agent. The vertical axis shows the change in $P(B_t = B^C)$ for the player-base. We see vendor agents with different motives will develop different behavioural characteristics, depending on the motives of the player-base. For example, nAch(1) vendors will tend to bargain with their customers, while nPow(1) vendors will tend to stick to their advertised prices. nAff(2) NPCs will often bargain, but will also exhibit a tendency to try to do whatever the player does. In some cases this results in a cycle between different strategies.

8.4   Battle of the Sexes Revisited

We can perform the same type of analysis for other abstract games when NPCs are to interact with multiple player characters. A battle of the sexes scenario, for example, abstracts certain kinds of interactions with quest-giving support charac-ters. As we saw in Chap. 7, the battle of the sexes game occurs when individuals have the choice of following their own goals, or adopting the goals of another. If both opt for their own goal, a (B^C, B^C) outcome results and each ends up going alone. The payoff for this outcome is $V^1 = V^2 = R$. A worse outcome, (B^D, B^D), results if both make the sacrifice of choosing the goal of the other, as they both end up alone engaging in a non-preferred activity. The payoff for this outcome is $V^1 = V^2 = P$. If, however, one chooses their own goal and the other plays the role of 'hero' and aids with that goal, then the outcome is better for both of them ($V^1 = T$ and $V^2 = S$ or $V^1 = S$ and $V^2 = T$). This description forms the basis for using the battle of the sexes game to model negotiation with quest givers.

8.4.1   Motivated Learning Agents as Quest Givers

Quest givers are NPCs that provide quests for players. A quest is a task given to a player character that yields a reward when completed. Quest givers may reward the

Fig. 8.5 Quest giver dialog
box. The player can make the
traditional 'accept quest' or
'reject-quest' choice, or can
start a mini-game negotiation
by proposing an alternative
quest for the agent to help
with

player with an item, points or money. Sometimes, quest givers are civilians scat-
tered throughout the game world, who offer advice or quests when the player
approaches them. *World of Warcraft* [4] and the *Elder Scrolls* [2] series employ this
strategy. Other games, like those in the *Final Fantasy* [1] series, will have NPCs
accompany the player as part of their 'party'.

Often, a player has the choice to either assist a quest giver or not. The NPC then
only follows the player if the player agrees to help with their quest. If NPCs are also
given the choice to help a player on one of their quests, however, then a battle of the
sexes scenario emerges. Figure 8.5 shows a dialog that permits a player to engage
in a negotiation with a quest giver NPC. The traditional 'accept' or 'reject' choices
are available (the 'I'm in, let's go!' and 'Not today thanks' buttons respectively).
We assume here that the 'I'm in, let's go!' button represents the B^D choices for the
player. A third button, labelled 'Help me with...', permits the player to select an
alternative quest with which the NPC should assist. This is a B^C choice by the
player and starts a mini-game that conforms to the format of a battle of the sexes
game. When the player clicks the 'Help me with...', a dialog box comes up from
which the player can select a different quest. When the player has selected the
alternative quest, the NPC is notified, but it is not immediately notified about which
quest has been proposed (i.e. the NPC does not have the advantage of knowing
whether the player has chosen B^C or B^D). The NPC then generates a B^C or B^D
response based on its own $P(B_t = B^C)$. Definitions of these behaviours are:

- B^D: agree to assist with player's proposed quest.
- B^C: continue to propose own quest (no change to the dialog box).

Finally, the NPC is notified of the player's decision. The possible outcomes are as
follows (the NPC choice is listed first in each case):

- A (B^D, B^D) outcome means the player and the NPC do not agree on a quest.
  The NPC's preferred quest is displayed on the dialog box.

- A (B^C, B^C) outcome means the player and the NPC do not agree on a quest
  (no change to the dialog box).

- A (B^C, B^D) outcome means the player and the NPC work to together on the NPC's quest. The player follows the NPC.

- A (B^D, B^C) outcome means the player and the NPC work together on the quest proposed by the player. The NPC follows the player.

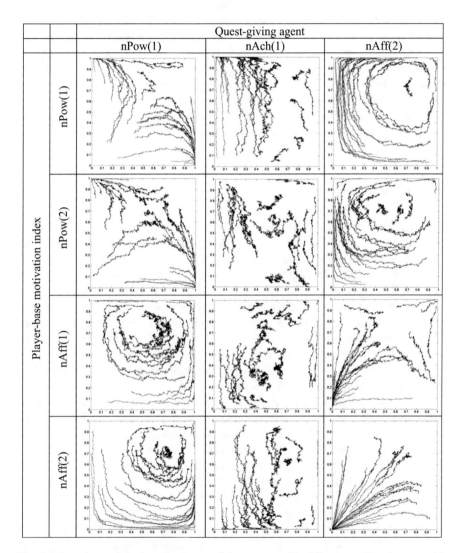

Fig. 8.6 Learning trajectories when quest-giving agents with different motives interact with player bases with different motive indices. The horizontal axis shows the change in $P(B_t = B^C)$ for the vendor agent. The vertical axis shows the change in $P(B_t = B^C)$ for the player base. See Chap. 6 for a description of the chart legend

8.4.2   *Empirical Study: Quest Givers Negotiating with Multiple Opponents*

Figure 8.6 shows the learning trajectories when quest giver agents with different motives interact with player-bases with different motive indices. The horizontal axis shows the change in $P(B_t = B^C)$ for the vendor agent. The vertical axis shows the change in $P(B_t = B^C)$ for the player-base. Again, we see that quest giver agents with different motives will develop different behavioural characteristics, depending on the motives of the player-base. nPow(1) quest givers will tend to agree to some sort of joint venture—either the NPCs quest or the player's quest—at least some of the time. Likewise, nAch(1) agents will tend to agree to some sort of joint venture, although they take longer to converge on a stable strategy than nPow(1) agents. nAff(2) agents prefer outcomes where both players adopt the same strategy.

8.5   Conclusion

In the last three chapters we have seen a number of ways in which an understanding of motivation theory can influence the design of NPCs in games. In the first instance, we have seen example applications of motivated agents for the control of NPCs and techniques for objective measurement of the resulting behavioural differences that occur between characters with different motives.

We have also seen that an understanding of motivation theory can be used to model the different types of responses that players might have to different strategies of play exhibited by NPCs. In Chaps. 6 and 7 we examined the different types of strategies that might emerge when NPCs with different motives encounter players with different motives. In this chapter we extended this idea to examine the possible influence and emergence of a player-base Zeitgeist.

We conclude that an understanding of motivation theory can give us novel techniques for the design of diverse NPCs, as well as a novel methodology for modelling the response of a player-base to such characters. In the case of NPCs, rule-based agents, learning agents and crowds of motivated agents can exhibit a diversity of behaviour. Learning agents can adapt their behaviour over time, before converging on a strategy that is best suited to the responses of their opponent.

This last point raises the question of what can be done if the learned strategy of a motivated agent becomes too easy for a player to predict. The next chapter addresses this question by considering an algorithm for the evolution of motivated agents so that the 'old generation' can be systematically replaced with a fresh set of agents.

Acknowledgments The *boids* implementation used as the basis of the *Breadcrumbs* game in Sect. 8.2 was based on code by Drew Heavner, downloaded from GitHub under an Apache License, Version 2.0. The Android wrapper was implemented by James Dannatt, a fourth year electrical engineering student at UNSW Canberra. Motivated agents were added by Kathryn Merrick.

References

1. Final Fantasy, *Square Enix* (1987)
2. The Elder Scrolls Series, *Bethesda Games Studio* (1994)
3. EverQuest, https://www.everquest.com. *Sony Online Entertainment* (1999)
4. T. Graepel, R. Herbrich, J. Gold, Learning to fight, in *Proceedings of the International Conference on Computer Games: Artificial Intelligence, Design and Education* (2004)
5. D. Heavner, r0adkll-flocking, Universal Java Flocking Engine. https://github.com/r0adkll (2013)
6. J. Laird, M. van Lent, Interactive computer games: human-level AI's killer application, in *Proceedings of the National Conference on Artificial Intelligence (AAAI)*, 2000, pp. 1171–1178 (2000)

Part IV
Evolution and the Future of Motivated Agents

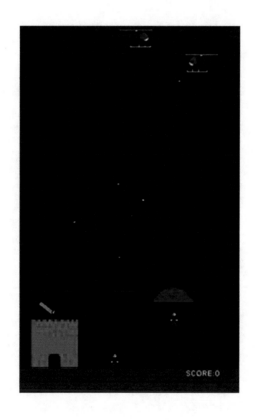

Chapter 9
Evolution of Motivated Agents

Part IV of this book looks to the future of motivated game-playing agents, first with a specific focus on the conditions under which agents with different motives might 'evolve' (this chapter) and then more broadly (Chap. 10). This chapter considers the evolution of motivation in a society of game-playing agents. Agents are studied first in a series of multiplayer social dilemma games. A framework is presented for the evolution of new generations of motivated agents when fitness is determined subjectively. We demonstrate how the composition of a society of motivated agents can change over time, and the evolutionary benefit of such change. We conclude with a study of the evolution of motivated agents in a game where their fitness is determined objectively and influenced by the skill of their human opponent.

9.1 The Evolutionary Perspective

Part III of this book studied motivated learning agents making decisions in mini-games and in-game scenarios that can be found in contemporary, popular computer game genres. This chapter and the next look, instead, towards some futuristic applications of motivated agents. First, this chapter develops a specific case study as an example: the evolution of artificial agents with different motives in a society of artificial agents. We consider the conditions under which individuals with different optimally motivating incentives might evolve, and discuss the contribution such agents might make to behavioural diversity in computer games. Chapter 10 then takes a broader look into the future of motivated agents.

In the previous chapter we saw that, over time, as motivated learning agents 'grow old' they begin to settle on a particular strategy that they have learned will satisfy their motives. As they become more predictable, they are more open to exploitation by player characters. One solution to this dilemma is to permit the 'old generation' to die out and be replaced with a new generation of characters. Characters in the new generation inherit aspects of the motive profiles of the most

© Springer International Publishing AG 2016

K.E. Merrick, *Computational Models of Motivation for Game-Playing Agents*, DOI 10.1007/978-3-319-33459-2_9

successful characters in the old generation. Success here can be measured either objectively or subjectively. An objective measure of success might include whether a particular non-player survives against a player. In contrast, in scenarios where the life force of a character is not threatened, subjectively successful agents might be those that can best satisfy their own motives. Both of these alternatives are examined in this chapter.

The remainder of this chapter is organised as follows. Section 9.2 examines the evolution of subjectively rational motivated agents. It uses four multiplayer social dilemma games as the setting. An empirical analysis is presented of motivated agents playing four n-player games: a common pool resource game, the n-player leader game, the hawk-dove game and an n-player battle of the sexes. These games are compound versions of the two-player games studied in Part III.

Finally, Sect. 9.3 returns to the *Paratrooper* game studied in Chap. 6 to examine the evolution of motivated agents in response to an objective fitness function.

9.2   Multiplayer Social Dilemma Games

As with the two-player games studied in Part III, in a multiplayer or 'n-player' social dilemma game, each of n players has two choices: B^C or B^D. The incentive for each choice depends on the combination of choices by all n players. An n-player game is mathematically equivalent to a compound two-player game in which each player plays one round against every other player and receives a payoff $1/(n{-}1)$ of that specified in the n-player game [1]. The total payoff of each player is the sum of the payoffs received in each two-player round. Thus, if we develop n-player games that are compounds of the two player games introduced in Chap. 6, then $R(n-1)$ is the reward if all players choose B^C, $P(n-1)$ is the punishment if all players choose B^D, $T(n-1)$ represents the temptation to choose B^D when the other players choose B^C and $S(n-1)$ is the sucker's payoff for choosing B^C when the other players choose B^D.

A number of specific compound multiplayer games can be defined by fixing the relationships between T, R, P and S:

- Common pool resource (CPR) games: $T > R > P > S$,

- n-player leader: $T > S > R > P$,

- The hawk-dove game: $T > R > S > P$ and

- n-player battle of the sexes: $S > T > R > P$.

The relative sizes of payoff values in these games follow the same orderings as those of the two-player prisoners' dilemma game, the leader game, the chicken game and the battle of the sexes game (see Chaps. 6 and 7). As with two-player games, we can visualize the n-player games in terms of their expected payoff, as

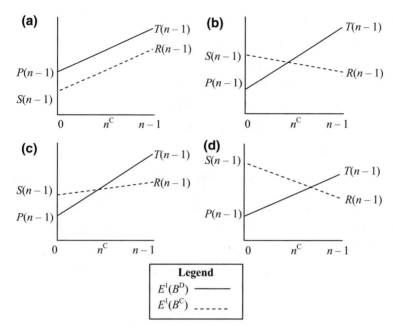

Fig. 9.1 Visualization of the payoff structures for a **a** common pool resource (CPR) game with $T > R > P > S$ **b** n-player leader game with $T > S > R > P$ **c** hawk-dove game with $T > R > S > P$ and **d** n-player battle of the sexes game with $S > T > R > P$

shown in Fig. 9.1. We can see from Fig. 9.1 that the payoff structure of each game is the same as the payoff structure of the corresponding two-player game (for example, compare Fig. 9.1a to Fig. 3.1). In fact, the two-player games are simply special cases of their n-player counterparts. We visualise the payoff of one player in a two-player game in terms of the probability of the other player choosing B^C. Although we visualise the payoff of a player in an n-player games in terms of the number of other players choosing B^C, this number is proportional to the probability of other players choosing B^C.

With the similarities of n-player games to their two-player counterparts in mind, it follows that transformations of the n-player games will produce the same orderings of subjective incentive values that resulted from the transformations of their two-player counterparts in Chaps. 6 and 7. It is thus unnecessary to present the game transformations again in this chapter, although examples of such transforms do exist [4]. Instead, this chapter will focus on studying empirical results of motivated evolutionary agents in n-player games. Algorithm 3.6 is used for the simulations in this section, with specific settings as follows:

- Line 1: The game world $\mathbf{W}$ (Eq. 9.1) is a 14×14 matrix comprised of subjective payoff functions for seven different types of agents. These are:

0. Rational agents; no motivation

1. nPow(1) agents: Strong power-motivated agents

2. nPow(2) agents: Weak power-motivated agents
3. nAch(1) agents: Achievement-motivated agents

4. nAch(2) agents: Achievement-motivated agents

5. nAff(1) agents: Weak affiliation-motivated agents

6. nAff(2) agents: Strong affiliation-motivated agents

The first type of agent uses the payoff of the game to compute evolutionary fitness. The latter six types of agents fall into each of the six categories examined theoretically in Chaps. 6 and 7. They use the game transformations presented in the following sections of this chapter to compute $\widehat{P}^k$, $\widehat{T}^k$, $\widehat{S}^k$ and $\widehat{R}^k$ for each agent type:

$$
\mathbf{W} = \begin{bmatrix}
P(n-1) & T(n-1) & \cdots & P(n-1) & T(n-1) \\
P(n-1) & R(n-1) & & S(n-1) & R(n-1) \\
\widehat{P}^1 & \widehat{T}^1 & \cdots & \widehat{P}^1 & \widehat{T}^1 \\
\widehat{S}^1 & \widehat{R}^1 & & \widehat{S}^1 & \widehat{R}^1 \\
& & \vdots & & \\
\widehat{P}^k & \widehat{T}^k & \cdots & \widehat{P}^k & \widehat{T}^k \\
\widehat{S}^k & \widehat{R}^k & & \widehat{S}^k & \widehat{R}^k \\
& & \vdots & & \\
\widehat{P}^6 & \widehat{T}^6 & \cdots & \widehat{P}^6 & \widehat{T}^6 \\
\widehat{S}^6 & \widehat{R}^6 & & \widehat{S}^6 & \widehat{R}^6
\end{bmatrix} \tag{9.1}
$$

- Line 2: We then construct a vector $\mathbf{x}_0$ stipulating the initial proportions of each type of agent choosing B^C and B^D (Eq. 9.2). Initially the society comprises only rational agents (without motivation). The simulation examines whether there is an evolutionary benefit of motivated agents by permitting mutation of motivated agents and allowing survival of agents with the highest fitness. For motivated agents, subjective incentive is used to measure fitness. For non-motivated agents, payoff is used to measure fitness.

$$\mathbf{x}_0 = \begin{bmatrix} 0.5 \\ 0.5 \\ 0 \\ 0 \\ 0 \\ 0 \\ 0 \\ 0 \\ 0 \\ 0 \\ 0 \\ 0 \\ 0 \\ 0 \end{bmatrix} \qquad (9.2)$$

- Line 2: Mutation is governed by the matrix $\mathbf{Q}$ shown in Eq. 9.3. Each type of agent has a probability of 0.98 of no mutations from within that type. Rational agents have a probability of 0.02 of mutating to nPow(1) agents, which themselves have an equal probability of choosing B^C or B^D. All the motivated agents have a probability of 0.02 of mutating to another type of agent with a slightly lower or slightly higher optimally motivating incentive (OMI).

$$\mathbf{Q} = \begin{bmatrix} 0.980 & 0.000 & 0.005 & 0.005 & \dots & 0.000 & 0.000 \\ 0.000 & 0.980 & 0.005 & 0.005 & & 0.000 & 0.000 \\ 0.010 & 0.010 & 0.980 & 0.000 & \dots & 0.000 & 0.000 \\ 0.010 & 0.010 & 0.000 & 0.980 & & 0.000 & 0.000 \\ 0.000 & 0.000 & 0.005 & 0.005 & \dots & 0.000 & 0.000 \\ 0.000 & 0.000 & 0.005 & 0.005 & & 0.000 & 0.000 \\ & & & \cdot & & & \\ & & & \cdot & & & \\ & & & \cdot & & & \\ 0.000 & 0.000 & 0.000 & 0.000 & \dots & 0.010 & 0.010 \\ 0.000 & 0.000 & 0.000 & 0.000 & & 0.010 & 0.010 \\ 0.000 & 0.000 & 0.000 & 0.000 & \dots & 0.980 & 0.000 \\ 0.000 & 0.000 & 0.000 & 0.000 & & 0.000 & 0.980 \end{bmatrix} \qquad (9.3)$$

- Line 3: $n = 101$.

- Line 7: $h = 0.001$.

- Line 8: $\mathbf{x}_{t+1}$ is normalised by subtracting from each element x^e the minimum value element $\min_e x^e$ and dividing each element of the resulting vector by $\sum_e x^e$.

No specific agent model is simulated between generations (line 5–6). The experiments in this chapter focus on the evolution of motives rather than on the impact a given motive has on behaviour between generations. Each simulation is run for 3,000 generations.

9.2.1 Common Pool Resource Games

In the real world, CPR games are concerned with 'common pool' resources such as energy or water. In a massively multiplayer online world a common pool resource might be a rare mineral such as 'mithril' or 'adamantite'. The dilemma confronting an individual citizen in a real or virtual world, when faced with a scarce or slow-to-regenerate resource, is whether to exercise restraint in resource gathering or use, or to exploit the restraint of others. An individual benefits from restraint only if a large proportion of others also exercise restraint. However, if this condition holds, it appears to become individually rational to ignore the call for restraint.

In the real world, some people exercise restraint while others exploit the restraint of others. Psychologists hypothesise that the asymmetric distribution of resources caused by different motivational preferences is an evolved mechanism that contributes to our survival as a species [2]. This chapter thus explores some computational conditions under which such a theory might hold.

The payoff structure for a CPR game is shown in Fig. 9.1a. This game is also known as an n-player prisoners' dilemma (nPD) because it uses the same assumption that $T > R > P > S$. Another in-game scenario that it can represent is thus a multiplayer arms race, as might occur in a turn-based strategy game.

The remainder of this section presents an empirical study of the evolution of 101 agents of seven different types playing a CPR game. We first discuss the experimental setup, which is common to all but one of the experiments in this chapter, and then the results.

9.2.1.1 Experimental Setup

First, we provide the game-specific implementation details of Algorithm 3.6. For the CPR game in this section, we use values $T = 4$, $R = 3$, $P = 2$ and $S = 1$. Specific OMI values for the different types of agents are:

0. Rational agents

1. nPow(1) agents: $\Omega^k = 375$

2. nPow(2) agents: $\Omega^k = 325$

3. nAch(1) agents: $\Omega^k = 275$

4. nAch(2) agents: $\Omega^k = 275$

5. nAff(1) agents: $\Omega^k = 175$

6. nAff(2) agents: $\Omega^k = 125$

This means that each type of agent will perceive the payoffs $T(n-1)$, $S(n-1)$, $R(n-1)$ and $P(n-1)$ differently. In a CPR game, the maximum payoff is $V^{max} = T(n-1)$, so the corresponding subjective incentive values for each payoff on substitution into Eq. 2.13 are:

$$\widehat{T}^k = T(n-1) - \left| T(n-1) - \Omega^k \right|, \tag{9.4}$$

$$\widehat{R}^k = T(n-1) - \left| R(n-1) - \Omega^k \right|, \tag{9.5}$$

$$\widehat{P}^k = T(n-1) - \left| P(n-1) - \Omega^k \right|, \tag{9.6}$$

$$\widehat{S}^k = T(n-1) - \left| S(n-1) - \Omega^k \right|, \tag{9.7}$$

where k denotes an agent type.

9.2.1.2   Metrics

Here, and in subsequent sections, we examine three charts that visualise the population dynamics that occur when different motivated and non-motivated agents (rational agents) interact in a multiplayer game. The first chart shows the fraction of

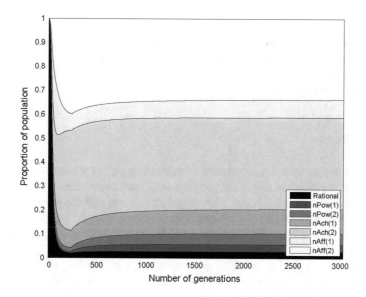

Fig. 9.2 Change in composition of a society of agents playing a common pool resource game

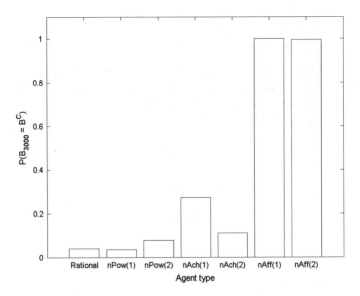

Fig. 9.3 Probability with which each type of agent chooses B^C by the end of the 3,000th generation of agents playing a common pool resource game. Affiliation-motivated agents are more likely to choose B^C while other agents are more likely to choose B^D

each type of agent in the population in each generation. That is, at each generation it plots for all agent types k

$$\sum_i F^k(B_t = B^i)$$

The second chart shows the probability with which agents of a given type will choose B^C at the end of the 3,000th generation, as defined in Eq. 3.23.

Finally, the third chart shows the subjective incentive perceived by each of the k types of motivated agent.

9.2.1.3 Results

Figure 9.2 shows that the proportion of rational agents drops dramatically in the first 100 generations as mutations progressively introduce different kinds of motivated agents into the population. We see that all of these mutants survive and thrive from generation to generation, but some form a greater ongoing proportion of the population than others. Specifically, nAch(2) agents and nAff(2) agents form approximately 65 % of the population by the end of the 3,000th generation. Figure 9.3 shows that the nAch(2) agents prefer the B^D choice, while the nAff(2) agents prefer the B^C choice. In fact, by the 3,000th generation 49 % of agents prefer the B^C choice and 51 % prefer the B^D choice.

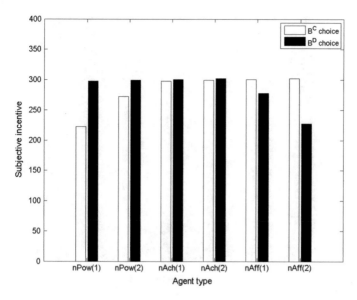

Fig. 9.4 Subjective incentive of each behaviour for different types of motivated agents playing a common pool resource game

Traditional, game-theoretic analysis of a society of rational agents predicts that 100 % of agents will prefer the B^D choice. The average payoff for all agents would be 200. However, in our simulation with motivated agents, the average payoff is higher: 248.6 on average. This is because some agents are subjectively fitter choosing B^C (see Fig. 9.4). This cooperation raises the objective fitness of the population. In this experiment we thus see an evolutionary benefit of motivation: differences in subjective fitness result in a diversity of agents and higher overall objective fitness of the society.

In the next section we examine a different scenario—the n-player leader game—in which there is no longer a clear evolutionary benefit of motivation, but population diversity is still achieved.

9.2.2   n-Player Leader

As we saw in Chap. 6, leader games emerge when individuals interact in scenarios when payoff is determined by the order in which they act. The highest payoff is awarded to those who act first. As with CPR games, multiplayer leader games can also model scenarios in which crowds of individuals compete for a resource. However, unlike CPR games, multiplayer leader games do not assume that the resource is scarce, only that reaching the resource first is preferable to waiting and reaching it later. This may be the case for a group of individuals moving towards a

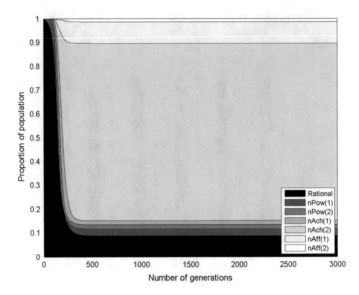

Fig. 9.5 Change in composition of a society of agents playing an *n*-player leader game

doorway, where those who hang back will have to wait longer to enter or exit. Likewise, it may be the case for a group of individuals moving to line up for service in a shop. Those who push forward will be served first. Those who hang back will still be served, but will have to wait longer.

9.2.2.1 Experimental Setup

The experiments in this section again use Algorithm 3.6. For the *n*-player leader game in this section, we use values $T = 4$, $R = 2$, $S = 3$ and $P = 1$. The game world **W** (Eq. 9.1) is a 14×14 matrix constructed using the same game transformations as those in Eqs. 9.4–9.7. The same seven types of agents are used.

9.2.2.2 Results

Figure 9.5 shows that the proportion of rational agents drops over the first 250 generations as mutations progressively introduce different kinds of motivated agents into the population. We see that all of these mutants again survive and thrive from generation to generation, but some form a greater ongoing proportion of the population than others. Different agents form the bulk of the population in this scenario than in the CPR scenario studied in the previous section. Specifically, Fig. 9.5 shows that nAch(2) agents form approximately 74 % of the population by the end of the 3,000th generation. nAff(1) agents are the next most prevalent at around 9 %.

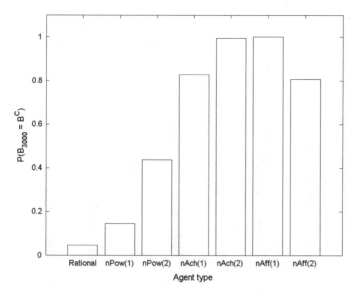

Fig. 9.6 Probability with which each type of agent chooses B^C by the end of the 3000th generation of an n-player leader game

Both the nAch(2) and the nAff(1) agents are predominantly B^C choosers (see Figs. 9.6 and 9.7). In fact, by the 3,000th generation 87 % of agents are B^C choosers and just 13 % are B^D choosers. At these proportions the average payoff of the society is 231.8, just under the average of 250 that would occur in a society of rational agents at equilibrium. In short, while a crowd of motivated agents might cooperate to share resources in a CPR game, they would still panic during an escape.

9.2.3 The Hawk-Dove Game

Like a CPR game, a hawk-dove game is yet another game that involves a contest over a sharable resource. Traditionally the resource in this case might be food or a mate. The contestants in the game are labelled as either 'hawks' or 'doves'. It should be noted that while in nature these are two different species that cannot cross-breed, the spirit of this game model is such that the terms 'hawk' and 'dove' refer to strategies used in the contest rather than to species of birds. The term hawk-dove was coined by Maynard Smith [3] during the Vietnam War when political views were polarised according to two prevalent views. The strategy of the 'hawk' (a fighter strategy) is to first display aggression, then escalate the confrontation into a fight until it either wins or is injured. The strategy of the 'dove' (fight avoider) is to first display aggression, but if faced with major escalation by an

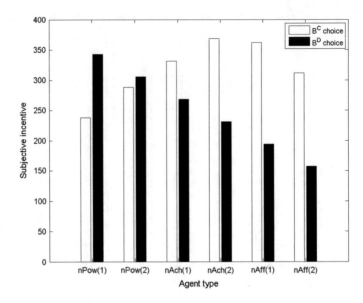

Fig. 9.7 Subjective incentive of each behaviour for different types of motivated agents playing an *n*-player leader game

opponent, to run for safety. If not faced with this level of escalation the dove will attempt to share the resource. We thus see that the scenario extends beyond animal conflict to other abstract scenarios.

The contested resource in a hawk-dove game is given the value U, and the damage from losing a fight is given a cost C. The cost of losing C is assumed to be greater than the value of winning U. Thus we have the following possibilities during pairwise interactions:

- A hawk meets a dove and the hawk gets the full resource. Thus $T = U$.

- A hawk meets another hawk of equal strength. Each wins half the time and loses half the time. Their average payoff is thus $P = \frac{U}{2} - \frac{C}{2}$ each.

- A dove meets a hawk. The dove backs off and gets nothing (that is, $S = 0$).

- A dove meets a dove and both share the resource ($R = \frac{U}{2}$ each).

As discussed previously, these payoffs are compounded to form an *n*-player game. That is, $\frac{(n-1)U}{2}$ is the reward if all players choose B^C, $\frac{(n-1)(U-C)}{2}$ is the punishment if all players choose B^D, $(n-1)U$ is the temptation to choose B^D when all other players choose B^C and 0 is the payoff for choosing B^C when the other players choose B^D.

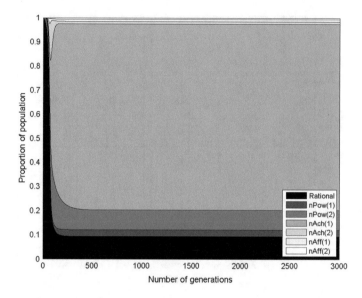

Fig. 9.8 Change in composition of a society of agents playing a hawk-dove game

9.2.3.1 Experimental Setup

For the hawk-dove game in this section, we use the assumption that $U = 4$ and $C = 8$. This gives us the non-compounded payoff values $T = 4$, $R = 2$, $S = 0$ and $P = -2$. The game world **W** (Eq. 9.1) is again a 14×14 matrix constructed using the transformations in Sect. 9.2.1 for seven different types of agents. In keeping with the raw payoff of the n-player hawk-dove game, these are:

0. Rational agents

1. nPow(1) agents: $\Omega^k = 350$

2. nPow(2) agents: $\Omega^k = 250$

3. nAch(1) agents: $\Omega^k = 150$

4. nAch(2) agents: $\Omega^k = 50$

5. nAff(1) agents: $\Omega^k = -50$

6. nAff(2) agents: $\Omega^k = -150$

9.2.3.2 Results

Figure 9.8 shows that the proportion of rational agents again drops over the first 100 generations as mutations progressively introduce different kinds of motivated

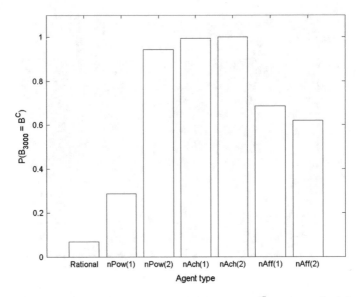

Fig. 9.9 Probability with which each type of agent chooses B^C by the end of the 3,000th generation of a hawk-dove game

agents into the population. In this scenario, we see that not all of these mutants survive and thrive from generation to generation. Some form significant, stable proportions of the population, but others do not increase in number beyond the 1 % mutation rate. B^C-choosing nAch(1) agents form the majority (almost 80 %) of the population by the 500th generation, and remain so by the 3,000th generation. Rational agents and nPow(2) agents are the next most prevalent agent types (see Fig. 9.8).

Traditional game-theoretic analysis of a society of rational agents predicts that 50 % of agents will choose B^D and 50 % B^C. The average payoff for all agents would be 100. However, in our simulation with motivated agents, the average payoff is higher: 194.1 on average. This is because more than 50 % of agents are subjectively fitter choosing B^C (see Figs. 9.9 and 9.10). This cooperation raises the objective fitness of the population. In this experiment we thus again see an evolutionary benefit of motivation, as well as some population diversity.

9.2.4 n-Player Battle of the Sexes

As we saw in Sect. 7.3, the battle of the sexes game models a predicament between friends with different interests and goals. Each prefers a certain goal that is different from the others' preferred goals, but all would rather stay together than act alone. If all opt to pursue their own goals (B^C), then all end up alone. The payoff for this

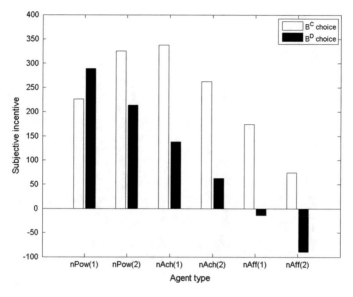

Fig. 9.10 Subjective incentive of each behaviour for different types of motivated agents playing a hawk-dove game

outcome is $(n-1)R$. A worse outcome results if all make the sacrifice of pursuing the goals of others (B^D), as they all end up alone pursuing those undesirable goals. The incentive for this outcome is $(n-1)P$. If, however, one agent chooses its preferred goal, while the others play the role of 'heroes' and makes the sacrifice of aiding with that goal, then the outcome for the agent whose goal is pursued is $S(n-1)$. The outcome for the heroes is $T(n-1)$.

9.2.4.1   Experimental Setup

The experiments in this section again use Algorithm 3.6. For the n-player battle of the sexes game in this section, we use values $T = 3$, $R = 2$, $S = 4$ and $P = 1$. The game world **W** (Eq. 9.1) is a 14×14 matrix. In a battle of the sexes game, the maximum payoff is $V^{\max} = S(n-1)$, so the corresponding subjective incentive values for each payoff on substitution into Eq. 2.13 are:

$$\widehat{T}^k = S(n-1) - \left| T(n-1) - \Omega^k \right|, \tag{9.8}$$

$$\widehat{R}^k = S(n-1) - \left| R(n-1) - \Omega^k \right|, \tag{9.9}$$

$$\widehat{P}^k = S(n-1) - \left| P(n-1) - \Omega^k \right|, \tag{9.10}$$

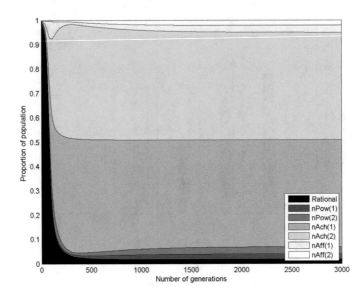

Fig. 9.11 Change in composition of a society of agents playing an *n*-player battle of the sexes game

$$\widehat{S}^k = S(n-1) - \left| S(n-1) - \Omega^k \right|, \tag{9.11}$$

where k denotes an agent type.

9.2.4.2 Results

Figure 9.11 shows that the proportion of rational agents drops over the first 250 generations as mutations progressively introduce different kinds of motivated agents into the population. We see that two of these mutants again survive and thrive from generation to generation, while others, including rational agents, do not survive at levels above the 1 % mutation rate.

Once again, different agents form the bulk of the population in this scenario than in the scenarios studied in the previous sections. Specifically, Fig. 9.11 shows that nAch(1) and nAch(2) agents combined form approximately 78 % of the population by the end of the 3,000th generation. nAch(1) agents are predominantly B^D choosers while nAch(2) are predominantly B^C choosers (see Figs. 9.12 and 9.13).

By the 3,000th generation the population is split almost evenly between B^D and B^C choosers (49:51). The average objective fitness of agents is 251.0, just over the average of 250 that would occur in a society of rational agents at equilibrium.

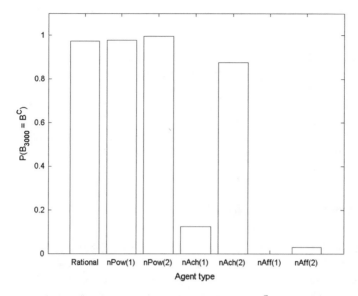

Fig. 9.12 Probability with which each type of agent chooses B^C by the end of the 3,000th generation of an n-player battle of the sexes game

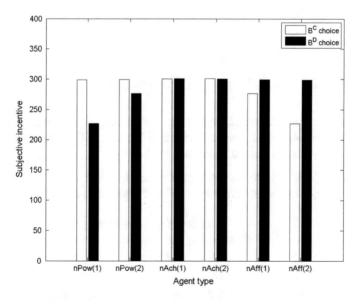

Fig. 9.13 Subjective incentive of each behaviour for different types of motivated agents playing an n-player battle of the sexes game

9.2.5   Summary

Our examination of evolution under conditions of subjective rationality reveals, perhaps most interestingly, that there is in many cases an evolutionary benefit of motivation. Specifically, heterogeneous societies can adapt, so that they achieve a greater objective fitness than a society of rational agents. Motivational diversity thus has its own intrinsic value.

The next section examines the complementary issue of the evolution of motivated agents in response to objective fitness only.

9.3   *Paratrooper* Revisited

The empirical studies above consider evolution of societies of non-player characters (NPCs) where fitness is determined subjectively by an agent's ability to satisfy its own motives. This section considers an application where fitness is determined objectively by an NPC's ability to survive. The two approaches are different, but complementary. When NPCs die regularly as part of a game (for example, enemies), objective fitness becomes increasingly important. In this section we return to our implementation of the *Paratrooper* game studied at the start of Chap. 6. Once again, the game's main antagonists, the paratroopers, are motivated agents. A majority of these agents die before they touch the ground, so objective fitness is critical.

9.3.1   Evolution of Motivated Paratroopers

In Chap. 6 we implemented motivated paratroopers using the dominant motive only method. In this section we use Algorithm 3.5 with an OMI representation of motivation. We use the specific settings as follows:

- Line 1: The objective fitness of a particular type k of agent is defined according to whether paratroopers of that type survive their descent or not.

$$f^k(x) = \begin{cases} 2 & \text{if } A^k \text{ lands safely} \\ 0 & \text{if } A^k \text{ is hit} \\ 1 & \text{otherwise} \end{cases} \tag{9.12}$$

- Line 2: $\mathbf{x}_0$ is initialised with the proportions of agents with different motives. We assumed 150 different OMI values in six categories:

1. nPow(1) agents: $\Omega^k \in [125, 149]$

2. nPow(2) agents $\Omega^k \in [100, 124]$

3. nAch(1) agents: $\Omega^k \in [75, 99]$

4. nAch(2) agents: $\Omega^k \in [50, 74]$

5. nAff(1) agents: $\Omega^k \in [25, 49]$

6. nAff(2) agents: $\Omega^k \in [0, 24]$

Initially, paratroopers are generated with equal probability ($Pr = 0.167$) of having an OMI in one of these categories. These OMI values fall in the range [0, 149] and are proportional to distance travelled before the opening of the parachute. Waiting longer to open the parachute has higher associated incentive, as the paratrooper falls faster and is thus in the air for less time and harder to hit.

- Line 3: The set **A** of agents is initially empty. During each 'tick' t of the game, multiple new paratroopers can drop from a helicopter (see Chap. 6 for a description of the game logic). The OMI of a new paratrooper is generated probabilistically from x_t.

- Line 6: The paratrooper will open its parachute when it reaches an altitude with an incentive that corresponds with its OMI (if it is still alive at that stage).

Evolution occurs dynamically during a game, and evolved population statistics are stored and used by the game controller in subsequent games.

9.3.2   Case Study

9.3.2.1   Experimental Setup

This section discusses a study of the evolution of motivated agents during 10 games (10 generations) against a single human player (the author in this study). Step size $h = 0.0003$. A small learning rate was chosen to prevent dramatic changes occurring in the playability of the game while it was in progress. In addition, no type of agent was permitted to exit the population entirely as a result of an evolutionary update. This permits an agent type to recover in response to a change in the strategy of the player.

9.3.2.2   Results

Figure 9.14 shows the change in fractions of agents with different OMIs over the course of the games. We see a modest increase in the fraction of power-motivated agents from 33.3 % during the first game to 37.2 % at the end of the 10th

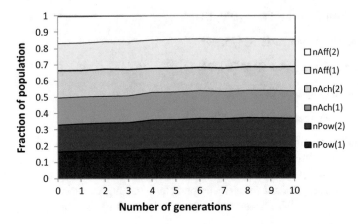

Fig. 9.14 Change in proportion of paratroopers with different motives, as a result of evolution in response to a survival-based, objective fitness function

generation (after 10 games). This reflects anecdotal evidence that it is more difficult to hit paratroopers that open their chutes at lower altitudes. Particularly, if the player is engaging paratroopers on one side of the screen, swivelling the turret to the other side of the screen to catch a free-falling paratrooper is difficult, although not impossible. As a result, paratroopers that have a higher OMI tend to land slightly more often.

As noted above, a small learning rate was chosen to prevent dramatic changes in the playability of the game while it was in progress. However, a larger learning rate would obviously mean that a greater change in the proportion of different types of agents would be observed over the course of the 10 generations.

9.4 Conclusion

The simulations and case study in this chapter have shown that if an evolutionary algorithm is used to create new generations of motivated agents in different scenarios:

- Different agents thrive in different scenarios;
- Diversity is achieved in all scenarios;
- When fitness is computed subjectively, an evolutionary benefit is sometimes observed in the form of higher average payoff achieved by the society;
- When fitness is computed objectively, societies of motivated agents can adapt the proportions of different types of agents to better suit conditions.

Evolutionary algorithms give us an alternative to characters that 'grow old' and become predictable. Such algorithms permit the 'old generation' to die out and be replaced with a new generation of characters that take on the most successful characteristics of the previous generation. This chapter has shown how inherited characteristics can include motivation. Results suggest that there may be quantifiable objective benefits in terms of performance when we permit such evolution to occur.

Acknowledgments The customised paratrooper game used in Sect. 9.3, and also depicted in the image at the start of Part IV, was implemented by Ben Quinton as part of a second year electrical engineering project at UNSW Canberra. Images used with permission. The evolutionary algorithm controlling the motivated agents was implemented by Kathryn Merrick.

References

1. A. Colman, *Game Theory and Experimental Games: The Study of Strategic Interaction* (Pergamon Press, Oxford, England, 1982)
2. J. Heckhausen, H. Heckhausen, *Motivation and Action* (Cambridge University Press, New York, NY, 2010)
3. J. Maynard-Smith, G.R. Price, The logic of animal conflict. Nature **246**, 15–18 (1973)
4. K. Merrick, Evolution of intrinsic motives in a multi-player common pool resource game, in *Proceedings of the IEEE Symposium Series on Computational Intelligence for Human-like Intelligence*, Orlando, Florida, 2014, pp. 36–43

Chapter 10
Conclusion and Future

Computational models of motivation have now been explored in a range of agent architectures. This book has focused on a set of four such architectures, but there is increasing recognition that motivation has a role to play in a range of different kinds of intelligent systems in the future, including computer games. This chapter concludes this book with a study of the different components of motivation that are currently available, the different agent architectures they can be used in and, finally, how these technologies may be used in future computer games.

10.1 The Building Blocks of Computational Motivation

'Motivated' [5, 15], 'self-motivated' [13] or 'intrinsically motivated' [20, 24] agents differ from agents without motivation by the inclusion of a module that can adapt the agent's focus of attention. The precise nature of the motivation module depends on the type of motivation it is modelling and the underlying agent model into which it is incorporated. However, some common components of computational models of motivation can be identified [15]. These include feature selection, goal generation, motivation, arbitration and goal-selection components.

This chapter concludes the book by placing the models developed in the preceding chapters in the context of these five components of computational motivation. This provides a methodology for using the models from this book in other settings. Specifically, this book has focused on the latter three components of motivation in the list above. Chapter 2 presented motivation functions for achievement, affiliation and power motivation. It showed how multiple motivations can be combined in a single profile (arbitration), and how motivation can be used for goal selection in either 'winner-takes-all' or probabilistic modes. Chapter 3 presented a number of agent algorithms that combine these components in different ways.

© Springer International Publishing AG 2016
K.E. Merrick, *Computational Models of Motivation for Game-Playing Agents*,
DOI 10.1007/978-3-319-33459-2_10

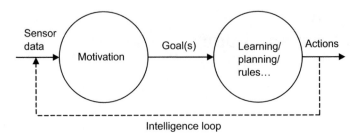

Fig. 10.1 The role of motivation in artificial agents

In the remainder of this chapter, we revisit the different combinations of motivation components that have been presented in the book, and consider other combinations that may be possible. We provide a mapping of motivated agent types to motivation functions as they are studied in this book in Sect. 10.2. Section 10.3 looks to the future development of computational motivation, including motivation functions themselves, the agent architectures in which they can be used and their application.

In the most general sense, computational models of motivation use sensor data as input and generate one or more goals to direct the activity of other reasoning processes, as shown in Fig. 10.1. These reasoning processes may be traditional learning, planning, crowd or evolutionary algorithms. The motivation module provides a dynamic input that is generated based on sensor data and the stored past experiences of the agent. As we outlined above, a number of common components of motivation are:

- Feature selection components;
- Task or goal generation components;
- Motivation functions;
- Arbitration functions;
- Goal-selection components;

The following sections summarise each component, and how the algorithms in this book might be integrated with each.

10.1.1   Feature Selection

Feature selection [6] algorithms focus attention on a subset of features (also called variables or attributes) extracted from data. In motivated agents, feature selection applied to sensor data does not necessarily imply a reduction in features as it does in traditional machine learning or statistics. Instead, it may be used to produce multiple different feature subsets—different views on sensor data—from which goal prototypes can be generated.

The need for feature selection in motivated agents has been debated, with a spectrum of approaches having been considered. Early approaches to subgoal generation in reinforcement learning—the precursors to computational motivation in this setting—assigned priority to features with different properties. For example, The HEXQ algorithm [9] focused attention on features according to the frequency with which their value changed. Later approaches recognised that multiple feature subsets could be constructed from the same input data [15] and advocated a lower degree of predefined rules for feature selection. Recent work has proposed algorithms that bypass traditional pre-processing of video data to perform adaptive feature selection [11].

Feature selection has not been the focus of the algorithms in this book, which are concerned with motivation of a set of known goals in constrained mini-games. However, the development of motivated agents that can adapt or respond in more open-ended settings will most likely need to consider the role of feature selection in motivation.

10.1.2   Task or Goal Generation

Recently there has been increased interest in designing agents for developmental robots that can bootstrap their own behaviour [27]. This includes agents that can choose which skills or knowledge to learn, as well as learn the chosen skills or knowledge. The broad implication is that a developmental agent must have the capacity to generate skill or knowledge acquisition goals autonomously as well as mechanisms that permit it to fulfil, or learn to fulfil, those goals. Accordingly, a number of recent computational frameworks for developmental learning [1, 2, 16, 17] have been designed with connected components for generating goals, prioritising goals using motivation functions and learning to fulfil goals.

Components for generating and prioritising goals have been proposed as extensions for a number of learning and optimisation algorithms, including active learning [3], reinforcement learning [15, 24] and particle swarm optimisation [7]. Approaches to the design of these components fall along a spectrum [12] defined by progressively stronger assumptions about the representation of goals. At one end of this spectrum lie models that direct learning by computing a scalar motivation value for each sensorimotor experience of a robot [10, 14, 16]. This value can be based on factors such as the level of resonance (or dissonance) [16] of new experiences with what is expected by the robot, or the novelty [14] or incentive [7] of new experiences. When combined with a reward-maximising learning or optimisation algorithm, motivation values have the effect of focusing the activities of the robot in a sub-region of the sensorimotor space. At the other end of the spectrum are approaches that formalise goals directly as sub-regions of some larger space [1, 2]. This is achieved by progressively splitting a goal space or task space into sub-regions. Other techniques such as competence progress or interest motivation can then be applied to prioritise goals.

Again, aside from a discussion of different types of goals in Chap. 1, goal generation has not been a focus of this book. Rather, we assume a set of goals and focus on algorithms (motivation functions) for prioritising these goals. However, goal generation is a critical aspect of computational motivation that is likely to receive a great deal of research attention in future.

10.1.3   Motivation Functions

A number of classifications of computational motivation functions have been proposed [18, 19]. One division classifies models as either knowledge- or competence-based. Knowledge-based models are hypothesised to play a greater role in exploratory behaviour, for example, to discover new goals. Competence-based models are hypothesised to play a greater role in deciding when to train skills to solve goals. In the category of knowledge-based models, computational curiosity [28] has been particularly well studied. Competence-based models on the other hand have been widely studied in learning settings.

One of the primary contributions of this book is a series of computational models of motivation that can be used for prioritising and selecting goals according to principles of power, achievement and affiliation motivation. These models take into account individuals' different preferences for certain kinds of incentives. Because we have not considered the design of feature selection and goal-generation modules in this book, the models sit outside of the knowledge- or competence-based classification scheme. Achievement motivation has potential for development into competence-based motivation. Alternatively, all three motives may be used in a knowledge-based manner (see [7], for example).

Regardless of how they are used, this book has proposed three approaches to the representation of power, achievement and affiliation motivation, listed here in order of decreasing complexity:

10.1.3.1   Profiles of Achievement, Affiliation and Power

Section 2.3.1 (Eq. 2.9) presents equations for profiles of achievement, affiliation and power motivation as a function of goal incentive. This model has 15 parameters controlling the shape of the motivation curve. The relative strengths of multiple motives can be taken into account in this model.

10.1.3.2   Modelling a Dominant Motive

Equation 2.10 of Sect. 2.3.2 presents a simplified approach in which only the dominant motive is modelled in the profile. This model has five parameters controlling the shape of the motivation curve.

10.1.3.3 Optimally Motivating Incentive

Section 2.3.3 presents a further simplified approach in which a motive profile is represented only by the value of the agent's optimally motivating incentive (OMI). This model requires storage of only a single scalar value.

10.1.4 Arbitration Functions

Arbitration functions are used to combine values from multiple motivation functions to produce a scalar motivation value [15]. This kind of combination occurs in the profiles of achievement, affiliation and power in Sect. 2.3.1. Multiple motivation curves are summed to produce a scalar motivation value.

Further consideration of arbitration functions is beyond the scope of this book. However, given the richness of motivation theories for humans and other natural systems, combination of multiple motivation functions in artificial agents is likely to be an area of active research in the future.

10.1.5 Goal Selection

Regardless of the way motivation is computed, the question remains as to what to do next. That is, once motivated, how should an artificial agent choose between highly motivating goals? As we have seen, in some motivated agent architectures—such as motivated reinforcement learning—motivation signals can be used directly as input for other reasoning processes. When other reasoning processes require more structured inputs—such as a goal structure—further processing is required. We presented two alternatives in Chap. 2, summarised briefly here.

10.1.5.1 Winner-Takes-All

In a 'winner-takes-all' approach (Sect. 2.4.1) the agent must pursue the goal with the highest motivation value. In the winner-takes-all approach the goal that a particular agent will pursue is chosen deterministically once motivation has been calculated. Behavioural variety can be achieved by having many agents with different motive profiles.

10.1.5.2 Probabilistic Goal Selection

In a probabilistic approach (Sect. 2.4.2) the agent has the greatest probability of pursuing the goal with the highest motivation, but may pursue goals of lesser

Table 10.1 Combinations of motivation components (motivation functions and goal-selection modes) studied in this book

Goal-selection mode	Motivation function		
	Optimally motivating incentive	Dominant motive only	Profile of achievement, affiliation and power
Winner-takes-all	Chap. 8	Chap. 4	Chap. 5
Probabilistic	Chaps. 6–8	Chap. 8	

motivation according to some probability distribution. This probability distribution may be informed by the difference in motivation values between goals. This means that if there are two goals with similarly high motivation values, there will be a similar probability that the agent will pursue either goal. Two agents with the same profile may make different goal selections as a result of this probabilistic approach, but over a long enough time period the proportion of times they will select a particular goal will be predictable.

10.1.6 Summary

To conclude this section, Table 10.1 summarises the combinations of motivation components that have been studied in this book, and where these studies can be found.

10.2 Motivated Agents

In addition to the different components of motivation that have been discussed in this book, we have also studied motivation in different agent architectures. The studies in this book expand the application of motivation in artificial systems beyond reinforcement and active learning systems, which have been the focus of development until now. Many of the agent architectures presented in this book have a much lower computational complexity than reinforcement and active learning, and can thus be applied to a larger number of agents, as might occur in a computer game. The following sections summarise the motivated agent architectures presented in this book, and also some of the existing architectures, to give a complete picture of the architectures currently available.

10.2.1 Motivated Rule-Based Agents

Introduced in Sect. 3.2.2, motivation in rule-based agents provides a way for agents to choose between different goals that are valid in a given state as a function of the incentive of the goal. Motivation is a meta-condition in motivated rule-based agents. Agents with different motive profiles (such as those studied in this book) or different experiences (for example, see [22]) will exhibit preferences for different goals in the same state.

10.2.2   Motivated Learning Agents

When agents can choose repeatedly from a set of goals and receive feedback on the incentive earned as a result of their choice, an opportunity for learning arises. A simple form of learning increases the agent's probability of selecting a particular behaviour in response to positive feedback, and decreases it otherwise.

Traditionally it has been assumed that all agents wish to maximise incentive earned. However, motivation theory suggests that behavioural diversity occurs because this assumption is inaccurate. As we saw Sect. 3.4.3, motivated learning agents model this weakened assumption using a built-in OMI and prefer behaviours that achieve incentives close to this value, regardless of whether or not they are maximal.

10.2.3   Motivated Reinforcement Learning Agents

Reinforcement learning agents [25] also learn from trial-and-error and reward feedback. However, a more complex decision-making environment is assumed, with behaviours causing the environment to transition to new states. Agents experiment with different behaviours in each state and receive reward or punishment from their environment. They progressively map reward earned to the states and/or behaviours that earn them and favour these states and/or behaviours over time.

When static reward signals are crafted around a particular goal, reinforcement learning agents will learn behaviours that solve that goal. However, with dynamic reward signals that use generic principles such as curiosity, motivated reinforcement learning (MRL) agents [15, 24] are created. MRL agents can focus their attention on different goals at different times. The agent designer no longer needs to know what goals the agent may encounter during its lifetime.

MRL agents were not considered in this book. However, this does not preclude future use of the motivation functions developed in this book within this framework.

10.2.4   Crowds of Motivated Agents

Motivated agents have been considered in rule-based 'boid' frameworks [22] and in multi-agent, swarm-based optimisation settings [7]. Different models of motivation have been used, including curiosity-based models in the first case, and achievement, affiliation and power models in the second. In this book, achievement, affiliation and power motivation have been embedded in 'boids' using the OMI approach (see Sect. 3.3.1). Motivation in these models has been used primarily to direct the exploratory behaviour of agents towards a wider range of targets than is traditionally possible with a greedy strategy.

10.2.5   Evolution of Motivated Agents

Finally, the evolution of motivated agents has been considered, also in different ways. The evolution of the proportions of different types of motivated agents in a population over successive generations was considered in Chap. 9 of this book. Chapter 5 and some previous work [23] have also considered the evolution of the parameters of motivation functions in individual agents to create agents that can match the decision-making performance of humans.

10.2.6   Summary

Table 10.2 summarises the different types of motivation that we considered in different agent architectures in this book. We group them according to the

Table 10.2 Combinations of motivation functions and motivated agent architectures studied in this book

	Agent architecture	→ Complexity →		
		Motivation function		
		Optimally motivating incentive	Dominant motive only	Profile of achievement, affiliation and power
Single agents	Motivated rule-based agents	Chap. 9	Chap. 4 Chap. 6	Chap. 5
	Motivated learning agents	Chap. 6 Chap. 7		
Multiple agents	Motivated crowds	Chap. 8		
	Evolution of motivated agents	Chap. 9		

complexity of the motivation function, and the applicability of the agent architecture to individual agents or groups of agents.

A majority of the work in this book has been along the axes of this table. That is, on the horizontal axis, we studied various motivation functions in motivated rule-based agents (the simplest architecture studied). Along the vertical axis we studied the OMI representation in all four of the agent architectures presented. Some work has been done in the literature exploring combinations in other regions of this table. For example, the evolution of agents with profiles of achievement, affiliation and power motivation has been studied previously [23]. However, in general the exploration of more combinations of motivation and agent technologies remains an open area of research.

10.3 Future

As the trend towards 'infinite resource' worlds continues, virtual worlds are becoming increasingly open-ended. They permit users to define their own content and control their own gaming experience. As this trend continues, there is an increasing requirement for new kinds of non-player characters (NPCs) to respond to user-created content. The roles for motivated agents, and the need for new computational motivation techniques will continue to grow.

10.3.1 Motivated Agents in Computer Games

In this book, we consider the primary gaming application of motivated agent technologies to be game artificial intelligence (game AI). This includes AI to control individual non-player characters, and groups of NPCs, as well as potentially strategic decision making in the game controller. We have already seen that NPCs can be further classified as enemies or opponents, pets and partner characters and support characters. Table 10.3 summarises the studies in this book by mapping

Table 10.3 Non-player character types and motivated agent architectures studied in this book

	Agent architecture	Non-player character type		
		Enemy or opponent	Partner or pet	Support
Single agents	Motivated rule-based agents	Chap. 6		
	Motivated learning agents	Chap. 6	Chap. 7	Chap. 8
Multiple agents	Motivated crowds			Chap. 8
	Evolution of motivated agents	Chap. 9		

these character types to the various algorithms studied. We see that there remains scope for application of the various agent architectures in other kinds of NPCs.

Outside of game AI, computational motivation also provides us with algorithms to assist with other advancing game technologies such as player experience modelling (PEM). PEM [21, 26] uses computational techniques to construct models of players' experiences and satisfaction levels. The aim is to permit future games to adapt dynamically to players' experiences, skill-sets, motivations and satisfaction levels to better maintain players in their 'flow zone' [4]. These technologies have potential for application outside of entertainment gaming and into 'serious game' applications such as simulation and training environments.

10.3.2 The Future of Computational Motivation

To conclude, the focus of this book has been on the so-called 'influential trio' of motives [8]: achievement, power and affiliation motivation. However, the field of motivational psychology recognises a much wider range of biological, cognitive and social motives that exist in humans [8]. Even within the narrow field of incentive-based motivation there can be complex motive profiles that combine different motives in different ways, resulting in different emergent behaviour. The relationship between motivation and leadership discussed in Chap. 1 (Sect. 1.3.5) is one such example.

Many of the motives and motive profiles that have been studied by motivational psychologists have the potential to be captured in computational models. This then opens the way for application to game-playing agents. Development of new models of motivation will continue to drive greater complexity and diversity of such game-playing agents.

The process of developing these models brings with it a number of challenges. The models need to be derived from psychological theories, and validated against either existing experimental data, or data collected in new experiments. These experiments need to be replicated in simulations with which artificial agents can then interact.

The study of computational motivation is likely to be an emerging area of research interest over the coming years as academics and developers alike appreciate the potential for greater autonomy and behavioural diversity among agents that have embedded motivation modules. As we have seen in this chapter, there remain some exciting research challenges to achieve these goals.

References

1. A. Baranes, P.-Y. Oudeyer, Intrinsically motivated goal exploration for active motor learning in robots: a case study, in *Proceedings of the IEEE/RSJ International Conference on Intelligent Robots and Systems*, pp. 1766–1773 (2010)
2. A. Baranes, P.-Y. Oudeyer, Maturationally-constrained competence-based intrinsically motivated learning, in *Proceedings of the IEEE International Conference on Development and Learning*, Ann Arbor, Michigan (2010)
3. A. Baranes, P.-Y. Oudeyer, Active learning of inverse models with intrinsically motivated goal exploration in robots. Robot. Auton. Syst. **61**, 49–73 (2012)
4. M. Csikszentmihalyi, *Flow: The Psychology of Optimal Experience* (Harper Perennial, New York, NY, 1990)
5. J. Graham, J. Starzyk, D. Jachyra, Opportunistic behavior in motivated learning agents. IEEE Trans. Neural Netw. Learning Syst. **26**, 1735–1746 (2014)
6. I. Guyon, A. Elisseeff, An introduction to variable and feature selection. J. Machine Learning Res. **3**, 1157–1182 (2003)
7. M. Hardhienata, K. Merrick, V. Ougrinovski, Task allocation in multi-agent systems using models of motivation and leadership, in *Proceedings of the IEEE Conference on Evolutionary Computation*, Brisbane, Australia, pp. 86–93 (2012)
8. J. Heckhausen, H. Heckhausen, *Motivation and Action* (Cambridge University Press, New York, NY, 2010)
9. B. Hengst, Discovering hierarchy in reinforcement learning with HEXQ, in *Proceedings of the Nineteenth International Conference on Machine Learning*, pp. 243–250 (2002)
10. F. Kaplan, P.-Y. Oudeyer, Motivational principles for visual know-how development, in *Proceedings of the Third International Workshop on Epigenetic Robotics: Modelling Cognitive Development in Robotic Systems, Lund University Cognitive Studies*, pp. 73–80 (2003)
11. V.R. Kompella, M. Luciw, J. Schmidhuber, Incremental slow feature analysis: adaptive low-complexity slow feature updating from high-dimensional input streams. Neural Comput. J. **24**, 2994–3024 (2012)
12. M. L. Maher, K. Merrick, B. Graham, Reasoning in the absence of goals, in *Proceedings of the AAAI Fall Symposium on Advances in Cognitive Systems* (2011)
13. J. Marshall, D. Blank, L. Meeden, An emergent framework for self-motivation in developmental robotics, in *Proceedings of the Third International Conference on Developmental Learning*, San Diego, CA, pp. 104–111 (2004)
14. S. Marsland, U. Nehmzow, J. Shapiro, Online novelty detection for autonomous mobile robots. J. Robot. Auton. Syst. **51**, 191–206 (2004)
15. K. Merrick, M.L. Maher, *Motivated Reinforcement Learning: Curious Characters for Multiuser Games* (Springer, Berlin, 2009)
16. K. Merrick, A comparative study of value systems for self-motivated exploration and learning by robots. IEEE Trans. Auton. Mental Dev. Spec. Issue Active Learning Intrinsically Motivated Exploration Robots **2**, 119–131 (2010)
17. K. Merrick, Modeling behavior cycles as a value system for developmental robots. Adapt. Behav. **18**, 237–257 (2010)
18. M. Mirolli, G. Baldassarre, Functions and mechanisms of intrinsic motivations: the knowledge versus competence distinction, in *Intrinsically Motivated Learning in Natural and Artificial Systems* (Springer, Berlin, 2012), pp. 49–72
19. P.-Y. Oudeyer, F. Kaplan, What is intrinsic motivation? A typology of computational approaches. Front. Neurorobot. **1** (2007)
20. P.-Y. Oudeyer, F. Kaplan, V.V. Hafner, Intrinsic motivation systems for autonomous mental development. IEEE Trans. Evol. Comput. **11**, 265–286 (2007)

21. C. Pedersen, J. Togelius, G. Yannakakis, Modeling player experience in Super Mario Bros., in *Proceedings of the IEEE Symposium on Computational Intelligence and Games*, Milan, Italy, pp. 132–139 (2009)
22. R. Saunders, J.S. Gero, Curious agents and situated design evaluations. Artif. Intell. Eng. Des. Anal. Manuf. **18**, 153–161 (2004)
23. K. Shafi, K. Merrick, E. Debie, Evolution of intrinsic motives in multi-agent simulations, in *Proceedings of the Ninth International Conference on Simulated Evolution and Learning*, pp. 198–207 (2012)
24. S. Singh, A.G. Barto, N. Chentanez, Intrinsically motivated reinforcement learning, in *Proceedings of the Advances in Neural Information Processing Systems 17 (NIPS)*, pp. 1281–1288 (2005)
25. R.S. Sutton, A.G. Barto, *Reinforcement Learning: An Introduction* (The MIT Press, 2000)
26. S. Tognetti, M. Garbarino, A. Bonarini, M. Matteucci, Modeling enjoyment preference from physiological responses in a car racing game, in *Proceedings of the IEEE Conference on Computational Intelligence and Games*, Copenhagen, Denmark, pp. 18–21 (2010)
27. J. Weng, J. McClelland, A. Pentland, O. Sporns, I. Stockman, M. Sur, E. Thelen, Artificial intelligence: autonomous mental development by robots and animals. Science **291**, 599–600 (2001)
28. Q. Wu, C. Miao, Curiosity: from psychology to computation. ACM Comput. Surveys **46** (2013)

Appendix A
Transformations of Social Dilemma Games

This appendix demonstrates the methodology for transforming a social dilemma game into a new game perceived by an agent with an optimally motivating incentive (OMI) Ω^j in a given range. This methodology was applied in Chaps. 6, 7 and 9 to produce transformations of the leader, chicken, prisoners' dilemma (PD) and battle of the sexes games. Section A.1 shows transformations of the PD game (with proofs) and Sects. A.2–A.4 show transformations of the leader, chicken and battle of the sexes games (proofs omitted). The basic process is the same for transforming other games.

A.1 Motivated Perception of the Prisoners' Dilemma

A.1.1 Power-Motivated Perception

For power-motivated agents, we assume $T > \Omega^j > 1/2(T+P)$. Two different games may be perceived by power-motivated agents, depending on whether Ω^j is closer to T or R. First, $T > \Omega^j > 1/2(T+R)$ (i.e. Ω^j closer to T) gives us the following transformation of the PD game using the equations in Chap. 6, Table 6.2 and simplifying the absolute values:

$$\widehat{T}^j = T - \left(T - \Omega^j\right) = \Omega^j \tag{A.1}$$

$$\widehat{R}^j = T - \left(\Omega^j - R\right) = T + R - \Omega^j \tag{A.2}$$

$$\widehat{P}^j = T - \left(\Omega^j - P\right) = T + P - \Omega^j \tag{A.3}$$

$$\widehat{S}^j = T - \left(\Omega^j - S\right) = T + S - \Omega^j \tag{A.4}$$

© Springer International Publishing AG 2016
K.E. Merrick, *Computational Models of Motivation for Game-Playing Agents*,
DOI 10.1007/978-3-319-33459-2

Theorem A.1 *When a player A^j with $T > \Omega^j > 1/2(T+R)$ perceives a PD game $\mathbf{W}$ with $T > R > P > S$, the game they perceive is still a valid PD with $\widehat{T}^j > \widehat{R}^j > \widehat{P}^j > \widehat{S}^j$.*

Proof If we assume $\widehat{R}^j \geq \widehat{T}^j$, then we have $T + R - \Omega^j \geq \Omega^j$, which simplifies to $1/2(T+R) \geq \Omega^j$. This contradicts the assumption that $T > \Omega^j > 1/2(T+R)$, so it must be true that $\widehat{T}^j > \widehat{R}^j$. If we assume that $\widehat{P}^j \geq \widehat{R}^j$, then we have $T + P - \Omega^j \geq T + R - \Omega^j$ or $P \geq R$, which contradicts the definition of the PD game. Thus, it must be true that $\widehat{R}^j > \widehat{P}^j$. Likewise, if we assume that $\widehat{S}^j \geq \widehat{P}^j$, then we have $T + S - \Omega^j \geq T + P - \Omega^j$, which simplifies to $S \geq P$. This also contradicts the definition of the PD game. Thus, it must be true that $\widehat{P}^j > \widehat{S}^j$. □

When $1/2(T+R) > \Omega^j > 1/2(T+P)$ (i.e. Ω^j closer to R), there is an additional transformation for the case when $\Omega^j < R$.

$$\widehat{R}^j = T - \left(R - \Omega^j\right) = T - R + \Omega^j \tag{A.5}$$

Theorem A.2 *When a player A^j with $1/2(T + R) > \Omega^j > 1/2(T + P)$ perceives a PD game $\mathbf{W}$ with $T > R > P > S$, the game they perceive will have $\widehat{R}^j > \widehat{T}^j > \widehat{P}^j > \widehat{S}^j$.*

Proof If we assume $\widehat{T}^j \geq \widehat{R}^j$, there are two possibilities. First, substituting Eqs. A.1 and A.2 gives $\Omega^j \geq T + R - \Omega^j$, which simplifies to $\Omega^j \geq 1/2(T+R)$. This contradicts the assumption that $1/2(T+R) > \Omega^j$. Similarly, substitution of Eqs. A.1 and A.5 gives $\Omega^j \geq T - R + \Omega^j$, which simplifies to $R \geq T$ and contradicts the definition of the PD game. Thus it must be true that $\widehat{R}^j > \widehat{T}^j$. If we assume that $\widehat{S}^j \geq \widehat{P}^j$ then we have $T + S - \Omega^j \geq T + P - \Omega^j$ which simplifies to $S \geq P$. This contradicts the definition of the PD. Thus, it must be true that $\widehat{P}^j > \widehat{S}^j$. Finally, if we assume $\widehat{P}^j > \widehat{T}^j$, we have $T + P - \Omega^j > \Omega^j$ or $1/2(T+P) > \Omega^j$, which contradicts the assumption for power-motivated agents that $\Omega^j > 1/2(T+P)$. Thus, it must be true that $\widehat{T}^j > \widehat{P}^j$. □

A.1.2 Achievement-Motivated Perception

For achievement-motivated individuals, we assume $1/2(T+P) > \Omega^j > 1/2(R+S)$. We again consider two cases, depending on whether Ω^j falls closer to R or P. When it falls closer to R, we have:

Theorem A.3 *When a player A^j with $1/2(T + P) > \Omega^j > 1/2(R + P)$ perceives a PD game* **W** *with $T > R > P > S$, the game they perceive is either:*
$$\widehat{R}^j > \widehat{P}^j > \widehat{T}^j > \widehat{S}^j \text{ if } \Omega^j > 1/2(T + S) \text{ or}$$
$$\widehat{R}^j > \widehat{P}^j > \widehat{S}^j > \widehat{T}^j \text{ if } \Omega^j < 1/2(T + S).$$

Proof $\widehat{R}^j > \widehat{T}^j$ and $\widehat{P}^j > \widehat{S}^j$ according to the proof of Theorem A.2. However, as we now assume $1/2(T + P) > \Omega^j$, we now have $\widehat{P}^j > \widehat{T}^j$ (converse of Theorem A.2). There are then two alternative orderings, differing in the position of $\widehat{S}^j$. If $\widehat{T}^j > \widehat{S}^j$, substitution of Eqs. A.1 and A.4 gives $\Omega^j > T + S - \Omega^j$, which simplifies to $\Omega^j > 1/2(T + S)$. Conversely if $\widehat{S}^j > \widehat{T}^j$, then $1/2(T + S) > \Omega^j$. $\square$

If Ω^j falls closer to P, the transformations in Eqs. A.1, A.5, A.3 and A.4 apply, with the addition of Eq. A.6 for the case when $\Omega^j < P$:

$$\widehat{P}^j = T - (P - \Omega^j) = T - P + \Omega^j \tag{A.6}$$

Theorem A.4 *When a player A^j with $1/2(R + P) > \Omega^j > 1/2(R + S)$ plays a PD game* **W** *with $T > R > P > S$, the game they perceive is either:*
$$\widehat{P}^j > \widehat{R}^j > \widehat{T}^j > \widehat{S}^j \text{ if } \Omega^j > 1/2(T + S) \text{ or}$$
$$\widehat{P}^j > \widehat{R}^j > \widehat{S}^j > \widehat{T}^j \text{ if } \Omega^j < 1/2(T + S)$$

Proof $\widehat{R}^j > \widehat{T}^j$ according to the proof of Theorem A.2. If we assume that $\widehat{S}^j \geq \widehat{P}^j$, substituting Eqs. A.4 and A.3 gives us $T + S - \Omega^j \geq T + P - \Omega^j$. This simplifies to $S \geq P$, which contradicts the definition of the PD. Likewise, substituting Eqs. A.4 and A.6 gives us $T + S - \Omega^j \geq T - P + \Omega^j$. This simplifies to $1/2(S + P) > \Omega^j$, which contradicts the assumption of an achievement-motivated agent. Thus it must be true that $\widehat{P}^j > \widehat{S}^j$. There are then two alternative orderings, differing in the position of $\widehat{S}^j$. If $\widehat{T}^j > \widehat{S}^j$, substitution of Eqs. A.1 and A.4 gives $\Omega^j > T + S - \Omega^j$, which simplifies to $\Omega^j > 1/2(T + S)$. Conversely if $\widehat{S}^j > \widehat{T}^j$, then $1/2(T + S) > \Omega^j$. $\square$

A.1.3 Affiliation-Motivated Perception

For affiliation-motivated agents, we assume $1/2(R + S) > \Omega^j > S$. We again consider two cases, depending on whether Ω^j is closer to P or S. First we consider the case where Ω^j is closer to P.

Theorem A.5 *When a player with* $1/2(R+S) > \Omega^j > 1/2(P+S)$ *plays a PD game* **W** *with* $T > R > P > S$, *the game they perceive has* $\widehat{P}^j > \widehat{S}^j > \widehat{R}^j > \widehat{T}^j$.

Proof If we assume $\widehat{T}^j \geq \widehat{R}^j$, substitution of Eqs. A.1 and A.5 gives us $\Omega^j \geq T - R + \Omega^j$, which simplifies to $R \geq T$. This contradicts the definition of the PD game, so it must be true that $\widehat{R}^j > \widehat{T}^j$. If we assume that $\widehat{S}^j \geq \widehat{P}^j$, substituting Eqs. A.4 and A.3 gives us $T + S - \Omega^j \geq T + P - \Omega^j$, which simplifies to $S \geq P$. This also contradicts the definition of the PD game. Likewise, substituting Eqs. A.4 and A.6 gives $T + S - \Omega^j \geq T - P + \Omega^j$, which simplifies to $1/2(P+S) \geq \Omega^j$ and contradicts the assumption that $\Omega^j > 1/2(P + S)$. Thus, it must be true that $\widehat{P}^j > \widehat{S}^j$. If we assume $\widehat{R}^j > \widehat{S}^j$, substitution of Eqs. A.5 and A.4 gives $T - R + \Omega^j > T + S - \Omega^j$. Simplification gives $\Omega^j > 1/2(R + S)$, which contradicts the assumption for affiliation-motivated agents that $1/2(R+S) > \Omega^j$. Thus, it must be true that $\widehat{S}^j > \widehat{R}^j$.□

If Ω^j is closer to S, then the transformations in Eqs. A.1, A.5, A.6 and A.4 apply.

Theorem A.6 *When a player with* $1/2(P + S) > \Omega^j > S$ *plays a PD game* **W** *with* $T > R > P > S$, *the game they perceive has* $\widehat{S}^j > \widehat{P}^j > \widehat{R}^j > \widehat{T}^j$.

Proof If we assume $\widehat{P}^j \geq \widehat{S}^j$, then we have $T - P + \Omega^j \geq T + S - \Omega^j$ which simplifies to $\Omega^j \geq 1/2(P + S)$. This contradicts the assumption that $1/2(P + S) > \Omega^j$. Thus, it must be true that $\widehat{S}^j > \widehat{P}^j$. If we assume $\widehat{R}^j \geq \widehat{P}^j$, then we have $T - R + \Omega^j \geq T - P + \Omega^j$, which simplifies to $P \geq R$. This contradicts the definition of the PD game. Thus, it must be true that $\widehat{P}^j > \widehat{R}^j$. Likewise, if we assume $\widehat{T}^j \geq \widehat{R}^j$, then we have $\Omega^j \geq T - R + \Omega^j$, which simplifies to $R \geq T$. This contradicts the definition of the PD game. Thus, it must be true that $\widehat{R}^j > \widehat{T}^j$□

A.2   Motivated Perception of the Leader Game

A.2.1   Power-Motivated Perception

For power-motivated agents, we assume $T > \Omega^j > 1/2(T + R)$. This gives us the same basic transformations as those in Eqs. A.1–A.4. There are two possible perceived games, depending on whether Ω^j is closer to T or S. When Ω^j is closer to T we have:

Theorem A.7 *When a player A^j with $T > \Omega^j > 1/2(T + S)$ perceives a Leader game* **W** *with $T > S > R > P$, the game they perceive is still a valid Leader game with $\widehat{T}^j > \widehat{S}^j > \widehat{R}^j > \widehat{P}^j$.*

When Ω^j is closer to S, an additional transformation is possible if $\Omega^j < S$:

$$\widehat{S}^j = T - \left(S - \Omega^j\right) = T - S + \Omega^j \tag{A.7}$$

A single perceived game still results as follows:

Theorem A.8 *When a player A^j with $1/2(T + S) > \Omega^j > 1/2(T + R)$ perceives a Leader game* **W** *with $T > S > R > P$, the game they perceive will have $\widehat{S}^j > \widehat{T}^j > \widehat{R}^j > \widehat{P}^j$. That is, the perceived game is a Battle of the Sexes game.*

A.2.2 Achievement-Motivated Perception

For achievement-motivated agents, we assume $1/2(T + R) > \Omega^j > 1/2(S + P)$. We again consider two cases, depending on whether Ω^j falls closer to S (Theorem A.9) or R (Theorem A.10). The transformations in Eqs. A.1–A.5 and A.7 are all required.

Theorem A.9 *When a player A^j with $1/2(T + R) > \Omega^j > 1/2(S + R)$ perceives a Leader game* **W** *with $T > S > R > P$, the game they perceive is either:*
$\widehat{S}^j > \widehat{R}^j > \widehat{T}^j > \widehat{P}^j$ *if* $\Omega^j > 1/2(T + P)$ *or*
$\widehat{S}^j > \widehat{R}^j > \widehat{P}^j > \widehat{T}^j$ *if* $\Omega^j < 1/2(T + P)$.

Theorem A.10 *When a player A^j with $1/2(S + R) > \Omega^j > 1/2(S + P)$ perceives a Leader game* **W** *with $T > S > R > P$, the game they perceive is either:*
$\widehat{R}^j > \widehat{S}^j > \widehat{T}^j > \widehat{P}^j$ *if* $\Omega^j > 1/2(T + P)$ *or*
$\widehat{R}^j > \widehat{S}^j > \widehat{P}^j > \widehat{T}^j$ *if* $\Omega^j < 1/2(T + P)$.

A.2.3 Affiliation-Motivated Perception

For affiliation-motivated agents, we assume $1/2(S + P) > \Omega^j > P$. There are again two cases depending on whether Ω^j is closer to R (Theorem A.11) or P (Theorem A.12). The transformations in Eqs. A.1–A.3, A.5 and A.7 are required.

Theorem A.11 *When a player A^j with $1/2(S + P) > \Omega^j > 1/2(R + P)$ perceives a Leader game* **W** *with $T > S > R > P$, the game they perceive is $\widehat{R}^j > \widehat{P}^j > \widehat{S}^j > \widehat{T}^j$.*

Theorem A.12 *When a player A^j with $1/2(R + P) > \Omega^j > P$ perceives a Leader game* **W** *with $T > S > R > P$, the game they perceive is $\widehat{P}^j > \widehat{R}^j > \widehat{S}^j > \widehat{T}^j$.*

A.3   Motivated Perception of the Chicken or Snowdrift Game

A.3.1   Power-Motivated Perception

For power-motivated agents playing a Chicken or Snowdrift game, we assume $T > \Omega^j > 1/2(T + S)$. This gives us the transformation in Eqs. A.1–A.4. There are two possible perceived games, depending on whether Ω^j is closer to T or R. When Ω^j is closer to T, we have:

Theorem A.13 *When a player A^j with $T > \Omega^j > 1/2(T + R)$ perceives a Chicken or Snowdrift game* **W** *with $T > R > S > P$, the game they perceive is still a valid Chicken or Snowdrift game with $\widehat{T}^j > \widehat{R}^j > \widehat{S}^j > \widehat{P}^j$.*

When Ω^j is closer to R, the additional transformation in Eq. A.5 is possible. A single perceived game still results as follows:

Theorem A.14 *When a player A^j with $1/2(T + R) > \Omega^j > 1/2(T + S)$ perceives a Chicken or Snowdrift game* **W** *with $T > R > S > P$, the game they perceive will have $\widehat{R}^j > \widehat{T}^j > \widehat{S}^j > \widehat{P}^j$.*

A.3.2   Achievement-Motivated Perception

For achievement motivated individuals we assume $1/2(T + S) > \Omega^j > 1/2(R + P)$. We again consider two cases, depending on whether Ω^j falls closer to R (Theorem A.15) or S (Theorem A.16). The transformations in Eqs. A.1–A.5 and A.7 are all required.

Theorem A.15 *When a player A^j with $1/2(T + S) > \Omega^j > 1/2(R + S)$ perceives a Chicken or Snowdrift game* **W** *with $T > R > S > P$, the game they perceive is either*
$$\widehat{R}^j > \widehat{S}^j > \widehat{T}^j > \widehat{P}^j \text{ if } \Omega^j > 1/2(T + P) \text{ or}$$
$$\widehat{R}^j > \widehat{S}^j > \widehat{P}^j > \widehat{T}^j \text{ if } \Omega^j < 1/2(T + P).$$

Theorem A.16 *When a player A^j with $1/2(R + S) > \Omega^j > 1/2(R + P)$ perceives a Chicken or Snowdrift game with $T > R > S > P$, the perceived game is either:*
$$\widehat{S}^j > \widehat{R}^j > \widehat{T}^j > \widehat{P}^j \text{ if } \Omega^j > 1/2(T + P) \text{ or}$$
$$\widehat{S}^j > \widehat{R}^j > \widehat{P}^j > \widehat{T}^j \text{ if } \Omega^j < 1/2(T + P).$$

A.3.3 Affiliation-Motivated Perception

For affiliation-motivated agents, we assume $1/2(R + P) > \Omega^j > P$. We consider two cases, depending on whether Ω^j is closer to S (Theorem A.17) or P (Theorem A.18). The additional transformation in Eq. A.7 is required for the case when $\Omega^j < S$.

Theorem A.17 *When a player A^j with $1/2(R + P) > \Omega^j > 1/2(P + S)$ perceives a Chicken or Snowdrift game with $T > R > S > P$, the perceived game is $\widehat{S}^j > \widehat{P}^j > \widehat{R}^j > \widehat{T}^j$.*

Theorem A.18 *When a player A^j with $1/2(S + P) > \Omega^j > P$ perceives a Chicken or Snowdrift game* **W** *with $T > R > S > P$, the game they perceive is $\widehat{P}^j > \widehat{S}^j > \widehat{R}^j > \widehat{T}^j$.*

A.4 Motivated Perception of the Battle of the Sexes Game

A.4.1 Power-Motivated Perception

In the battle of the sexes game, the maximum incentive is $V^{\text{max}} = S$. This gives us a new set of transformations upon substitution into Eq. 2.13:

$$\widehat{T}^j = S - (\Omega^j - T) = S + T - \Omega^j \tag{A.8}$$

$$\widehat{R}^j = S - \left(\Omega^j - R\right) = S + R - \Omega^j \tag{A.9}$$

$$\widehat{P}^j = S - \left(\Omega^j - P\right) = S + P - \Omega^j \tag{A.10}$$

$$\widehat{S}^j = S - \left(S - \Omega^j\right) = \Omega^j \tag{A.11}$$

For power-motivated agents, we assume $S > \Omega^j > 1/2(S + R)$. There are two possible perceived games, depending on whether Ω^j is closer to T or S.

Theorem A.19 *When a player A^j with $S > \Omega^j > 1/2(S + T)$ perceives a Battle of the Sexes game* **W** *with $S > T > R > P$, the game they perceive is still a valid Battle of the Sexes game with $\widehat{S}^j > \widehat{T}^j > \widehat{R}^j > \widehat{P}^j$.*

There is an additional transformation when $T > \Omega^j$:

$$\widehat{T}^j = S - \left(T - \Omega^j\right) = S - T + \Omega^j \tag{A.12}$$

Theorem A.20 *When a player A^j with $1/2(S + T) > \Omega^j > 1/2(S + R)$ perceives a Battle of the Sexes game* **W** *with $S > T > R > P$, the game they perceive will have $\widehat{T}^j > \widehat{S}^j > \widehat{R}^j > \widehat{P}^j$. That is, the perceived game is a Leader game.*

A.4.2 Achievement-Motivated Perception

For achievement-motivated agents, we assume $1/2(S + R) > \Omega^j > 1/2(T + P)$. We again consider two cases, depending on whether Ω^j falls closer to T (Theorem A.21) or R (Theorem A.22):

Theorem A.21 *When a player A^j with $1/2(S + R) > \Omega^j > 1/2(T + R)$ perceives a Battle of the Sexes game* **W** *with $S > T > R > P$, the game they perceive is either:*
$\widehat{T}^j > \widehat{R}^j > \widehat{S}^j > \widehat{P}^j$ *if* $\Omega^j > 1/2(S + P)$ *or*
$\widehat{T}^j > \widehat{R}^j > \widehat{P}^j > \widehat{S}^j$ *if* $\Omega^j < 1/2(S + P)$.

There is an additional transformation when $R > \Omega^j$

$$\widehat{R}^j = S - \left(R - \Omega^j\right) = S - R + \Omega^j \tag{A.13}$$

Theorem A.22 *When a player A^j with $1/2(T + R) > \Omega^j > 1/2(T + P)$ perceives a Battle of the Sexes game* **W** *with $S > T > R > P$, the game they perceive is either:*
$\widehat{R}^j > \widehat{T}^j > \widehat{S}^j > \widehat{P}^j$ *if* $\Omega^j > 1/2(S + P)$ *or*
$\widehat{R}^j > \widehat{T}^j > \widehat{P}^j > \widehat{S}^j$ *if* $\Omega^j < 1/2(S + P)$.

A.4.3 Affiliation-Motivated Perception

For affiliation-motivated agents, we assume $1/2(T + P) > \Omega^j > P$. There are again two cases, depending on whether Ω^j is closer to R (Theorem A.23) or P (Theorem A.24).

Theorem A.23 *When a player A^j with $1/2(T + P) > \Omega^j > 1/2(R + P)$ perceives a Battle of the Sexes game* **W** *with $S > T > R > P$, the game they perceive is $\widehat{R}^j > \widehat{P}^j > \widehat{T}^j > \widehat{S}^j$.*

Theorem A.24 *When a player A^j with $1/2(R + P) > \Omega^j > P$ perceives a Battle of the Sexes game* **W** *with $S > T > R > P$, the game they perceive is $\widehat{P}^j > \widehat{R}^j > \widehat{T}^j > \widehat{S}^j$.*

Appendix B
Flocking Parameter Values
Used in *Breadcrumbs*

Number of boids:

```
private static final int BOID_AMOUNT = 250;
```

World size:

```
public static final int WIDTH = 640;
public static final int HEIGHT = 480;
```

Rule weightings, constraints and radii of application:

```
engine.setAlignment(0.80f);
engine.setCohesion(0.65f);
engine.setSeparation(0.60f);

engine.setInteractionRadius(18);
engine.setSeparationRadius(12);
engine.setInnerSeperationRadius(7);

private static final float cohesionMin = -0.1f;
private static final float cohesionMax = 5f;
private static final float alignmentMin = -0.01f;
private static final float alignmentMax = 0.05f;
private static final float separationMin = 6f;
private static final float separationMax = 7f;

float radius = 20f;
float sepRadius = 12f;
float innerSepRadius = 7;
```

Velocity settings:

```
engine.setSpeedLimit(80);
new Vector2D(MathUtils.random(-20, 20) // initial velocities
```

© Springer International Publishing AG 2016   207
K.E. Merrick, *Computational Models of Motivation for Game-Playing Agents*,
DOI 10.1007/978-3-319-33459-2

Index

A

Achievement, 7
Achievement motivation, 10
  mastery-oriented, 18
  modelling, 26, 72
  performance-oriented, 18
  with learning, model of, 79
Advancement, 7, 17
Affiliation, v, 13
Affiliation motivation, 13
  modelling, 31
Agents
  H-H AchAgent, 73, 75
  H-L AchAgents, 73, 76
  L-H AchAgents, 73, 76, 80
  L-L AchAgents, 73
  motivated reinforcement learning, 53, 189
  motivated rule-based, vi, vii, 48, 188
    prisoners' dilemma and, 92
  motivated rule-based, as paratroopers, 101
  motivated rule-based, game scenarios, 47
  motivated rule-based, modelling
    achievement, 72
  motivated rule-based, playing roulette, 87
  nAch(1), 107, 111, 113–115, 117–119,
    129, 131, 132, 138, 141, 154, 175
  nAch(2), 107, 114, 115, 129, 138, 165, 169,
    172, 177, 180
  nAchAgents, 85, 86, 88
  nAchCompAgent, 120, 121, 123, 133, 135
  nAchCoopAgent, 121, 124, 134, 135
  nAchNeutralAgent, 120, 121, 124, 133,
    135, 136
  nAff(1), 107, 115, 129, 138, 155, 166, 168,
    172

  nAff(2), 107, 111–113, 115, 117–119, 129,
    131, 132, 138, 140–142, 166, 169
  nAffAgent, 85, 86, 88, 92
  nAffNeutralAgent, 121, 124, 134, 136
  nPow(1), 107, 111, 113–115, 117–119,
    129, 131, 138, 140–142, 154, 155,
    158, 167
  nPow(2), 107, 108, 114, 129, 138, 155,
    165, 168, 175
  nPowAgents, 85, 88, 92, 94
  nPowNeutralAgent, 120, 121, 123, 124,
    135
  rational, 165, 166, 169–172, 175, 177
  RTMAgents, 78
  rule-based, v, vi, 46, 47
  SimAgents, 79
Algorithm, 49, 52, 58, 63
Approach-avoidance conflict, 13, 22
Approach-avoidance motivation, 12, 21, 22
Approach-avoidance theory, 22
Approach goal, 10, 11
  mastery, 11
  performance, 11
Arbitration, 183, 187
Arms race, 54, 56, 106, 110
  in turn-based strategy games, 106
Avoidance goal, 10, 11
  mastery, 11
  performance, 11

B

Bargaining, 153
Battle of the sexes, 137, 155
  motivated perception of, 201
  n-player, 175

© Springer International Publishing AG 2016
K.E. Merrick, *Computational Models of Motivation for Game-Playing Agents*,
DOI 10.1007/978-3-319-33459-2

perception by motivated learning agents,
        138
Battle pet, 127, 133
Boids, 49, 51, 146, 147
Breadcrumbs, 145–148, 150, 205

C
Character
    battle pets, 127
    competitor, 99, 100
    enemy, 47, 48, 101, 133, 134, 136
    minion, 127, 133
    non-player, v–viii, 3, 4, 100, 127, 128, 164
    non-player, artificial intelligence in, 45
    non-player, game-playing agents and, 45
    non-player, types of, 99
    opponent, 99, 114, 121, 153, 158
    partner, vii, 99, 100, 129
    partner, motivated learning agents and, 131
    pets, 127
    player, vii, 3, 4, 7, 8, 46, 47, 70, 84, 127,
        138, 155, 163, 170
    quest giver, 145, 155, 158
    support, 49, 99, 100, 145, 151, 152, 155,
        191
    vendor, 152, 154
Coercive power, 14
Common pool resource (CPR), 164, 167
Common-pool resource game. *see* Game:
        Common-pool resource
Competence, 11, 14, 18, 185
Competence-based motivation, 186
Competitor, 99, 100
Conflict
    approach-avoidance, 11, 13, 22
Crowd
    greedy, 148
    motivated, vii, 51, 145, 148
    random, 148, 151
Curiosity, 18, 26, 186, 190
Customisation, 6, 7

D
Dominant motive, 15, 16, 34–36, 86, 107, 186

E
Enemy, 47, 48, 101, 105, 133, 134, 136
Escapism, 7
Evolution, 59, 164
    in multiplayer social dilemma games, 60
    of motivated agents, 60, 163, 190

of motivated agents using objective fitness,
        61
of motivated agents using subjective fitness,
        61
of motivated paratroopers, 180
Expert power, 14
Explicit incentives, 9, 38, 39, 64

F
Failure
    incentive to avoid, 10, 11
    motivation, 11, 73
    probability of, 10
Fear, 13, 31, 32, 90
Feature selection, 184
Finite-resource worlds, 6
Fitness
    objective, 61, 164, 175, 177, 182
    subjective, 61, 163
Flocking, 49, 51, 147, 148
Function
    arbitration, 35, 184, 187
    fitness, 60, 164
    Gaussian, 25
    hyperbolic, 22
    linear, 22
    motivation, 39, 185, 186, 190, 191
    quadratic, 23
    sigmoid, 25–27, 31, 80

G
Game
    battle of the sexes, 128, 137, 138, 140, 141,
        155, 156, 164, 175, 195, 201
    Breadcrumbs, 145–148, 150, 151, 205
    chicken, 128, 151, 152, 200
    common-pool resource, 164, 167
    hawk-dove, 164, 172, 174
    leader, 113, 115, 117, 138
    mini, 54, 69, 100, 153, 163
    mixed-motive, 100
    multiplayer, 4, 59, 163
    n-player, 59, 60, 164, 165, 170, 171
    n-player battle of the sexes, 164, 175, 177
    n-player leader, 170
    online, 3
    Paratrooper, 100, 101, 103, 179
    performance-approach, 11
    prisoners' dilemma, 90, 92, 105
    quoits, 70
    ring-toss, 70, 75

roulette, 84, 87
roulette, motive profiles for, 85
serious, 3, 192
snowdrift, 128, 129, 131, 151
social dilemma, 53, 54, 56, 59, 60, 106, 164, 195
turn-based strategy, 54, 106, 114, 119
Game data mining, v, 3, 4, 45, 46
Game theory, 55, 56, 60, 83, 90, 92, 106
Game world, 4, 5, 17, 46, 69, 145, 156
Gaussian function, 25
Goal
    avoidance, 10, 11
    mastery-approach, 11
    mastery-avoidance, 11
    performance-avoidance, 11
Goal selection
    probabilistic, 41, 48, 102, 147, 183, 187
    winner-takes-all, 40, 48, 183, 187
Gradient of approach, 13, 22, 23, 30
Gradient of avoidance, 13, 30, 33, 75
Greedy crowds, 148
Griefers, 8, 17

H
Hawk-dove game, 164, 172, 174
H-H AchAgent, 72, 73, 75, 80
H-L AchAgent, 71, 72
Hope, 13, 31, 32
Hyperbola, 22

I
Immersion, 6, 7
Imperial motive profile, 15, 35
Implicit incentive, 9
Implicit motivation, 21, 39
Implicit motive profile, 15
Incentive, 9
    explicit, 9, 38, 39, 57, 64
    for success, 10, 33
    implicit, 9, 15, 39
    optimally motivating, 36, 106, 147, 163, 187, 195
    subjective, 38, 39, 41, 57, 63, 107, 165, 177
    to avoid failure, 10, 11
Informational power, 14
Inhibition, 14
Interest, 4
Interest graph, 5, 7, 18
Intimacy, 13

K
Killers, 5, 17
Knowledge-based motivation, 186

L
Leader game, 114, 115, 117, 119, 138, 164, 170
    motivated perception of, 198
Leadership motive profile (LMP), 15
Learning agents, 52
    as explorers, 117
    as negotiators, 140
    as partner characters, 131
    as quest givers, 155
    as vendors, 152
    motivated, 56, 189
    motivated reinforcement, 189
Legitimate power, 14
L-H AchAgent, 73, 76, 80
Linear function, 103
L-L AchAgent, 72, 73, 75, 76, 80

M
Mandler-Sarason test of test anxiety, 29, 71
Mastery-approach goal, 11
Mastery-avoidance goal, 11
Mastery-oriented achievement motivation, 18, 80
Mastery-oriented estimate, 29
Mini-game, 5, 54
Minion, 127, 133
Mixed-motive game, 100
Motivated crowds, 51, 148
Motivated learning agents, vi, 46, 53, 56, 106, 129, 138, 189 *See also under* Agents
    algorithm, 58
    as explorers, 117
    as negotiators, 140
    as partner characters, 131
    as vendors, 152
    in arms race, 110
    quest givers, 155
Motivated rule-based agent, vi, vii, 48, 64, 72, 92, 94, 100, 101, 104, 188
    as paratroopers, 101
    game scenarios for, 47
    playing roulette, 87
Motivation
    achievement. *See* Achievement motivation
    affiliation. *See* Affiliation motivation

Motivation (*cont.*)
    approach-avoidance. *See*
    Approach-avoidance motivation
    curiosity, 18, 189, 190
    intimacy, 13
    power, vi, 14, 21, 83
    power, modelling, 33
    success, 10, 11
Motivation, approach-avoidance, 12, 21, 22,
    27, 30, 33, 36
Motive
    dominant. *See* Dominant motive
    implicit. *See* Implicit motive
Motive profile
    imperial. *See* Imperial motive profile
    implicit. *See* Implicit motive profile
    leadership. *See* Leadership motive profile
    (LMP)
Multiplayer game, 164, 169
Multiplayer social dilemma games, 59, 164

N
nAch(1). *See under* Agents
nAch(2). *See under* Agents
nAchAgent. *See under* Agents
nAchCompAgent. *See under* Agents
nAchCoopAgent. *See under* Agents
nAchNeutralAgent. *See under* Agents
nAff(1). *See under* Agents
nAff(2). *See under* Agents
nAffAgent. *See under* Agents
nAffNeutralAgent. *See under* Agents
Nash equilibrium (NE), 55, 111
National motive index (NMI), 16, 151
Negotiation, 138, 156
Non-player character. *See under* Character
Norm-based estimate, 29
Novelty, 18, 26
n-player game, 164
n-player leader game, 170
n-player social dilemma game, 59
nPow(1). *See under* Agents
NPow(2). *See under* Agents
nPowAgent. *See under* Agents
nPowNeutralAgent. *See under* Agents

O
Objective fitness. *See under* Fitness
Online games, v, 3, 4
Opponent character. *See under* Character
Optimally motivating incentive (OMI), 36, 41,
    106, 129

P
Pacifist, 7
Paratrooper game, 100, 101, 103, 164, 180,
    179
Partner. *See under* Character
Payoff, 53, 55, 57, 59, 91, 92, 105, 113, 128,
    137, 164, 170, 175
Perceived incentive, 39
Performance-approach goals, 11
Performance-avoidance goals, 11
Performance-oriented achievement motivation.
    See under Achievement motivation
Personalised power, 4
Pets, 127, 133, 191
Player. *See under* Character
Player-base motive index, 146, 151
Player experience modelling, 3, 4, 45, 192
Player type
    achiever, 4, 6
    explorer, 4–6, 18
    griefer, 8, 17
    killer, 4–6, 8, 17
    pacifist, 7
    runner, 7
    socialiser, 4–6
    solver, 7
    veteran, 7
Player-versus-environment (PVE), 17
Player-versus-player (PVP), 17
Power
    coercive. *See* Coercive power
    expert. *See* Expert power
    informational. *See* Informational power
    legitimate, 14
    motivation, vi, 14–16
    personalised. *See* Personalised power
    referent. *See* Referent power
    reward. *See* Reward power
Prisoners' dilemma game, 83, 90
    motivated perception of, 195
Probabilistic goal selection, 41, 48, 187
Probability of failure. *See under* Failure
Probability of success, 10, 11, 13, 14, 27–29,
    31, 32, 39, 47, 72, 73, 80, 85
Procedural content generation, 45
Projective measure of need achievement, 70

Q
Quadratic function. *See* Function: Quadratic
Quest giver. *See* Character: Quest giver
Quoits, 70

R
Random crowds, 148
Rational agent. *See* Agent: Rational
Referent power, 14
Reinforcement learning (RL), 53, 58, 79
Rejection, 13, 31
Reward power, 14
Ring-toss game, vii, 70, 75, 81
Risk-taking model (RTM), 10, 26, 77
Role-playing game, 4, 6
Roulette game, 84
RTMAgent. *See* Agents: RTMAgent
Rule-based agent. *See* Agent: Rule-based
Runner. *See* Player-type: Runner

S
Self-based estimates, 27, 29
Serious game, 3, 192
Sigmoid. *See* Function: Sigmoid
SimAgent. *See* Agent: SimAgent
Snowdrift game, 129, 131, 200
Social comparison standards, 27, 29
Social dilemma game, viii, 53, 56, 59, 60, 90,
        106, 164, 195
Socialiser. *See* Player-type: Socialiser
Solver. *See* Player-type: Solver
Strategy, 55, 106
Subjective incentive, 38, 39, 41, 57, 63, 107,
        165, 177
Subjective rationality, 57, 179
Success
    incentive for, 10, 33, 36
    motivation, 10, 11

    probability of, 10, 11, 13, 14, 24, 27–29,
        31, 39
Support character. *See* Character: Support

T
Task, 10, 52
Task-based estimates, 29
Thematic apperception test (TAT), 70
Three factor theory, 9
Three needs theory, 9
Turn-based strategy (TBS) game, 54, 106, 114,
        119
Turning point, 26, 27

V
Vendor. *See* Character: Vendor
Veteran. *See* Player-type: Veteran
Virtual world, v, viii, 3, 4, 9, 45, 56, 150, 191

W
Winner-takes-all, 40, 48, 187
    goal selection. *See* Goal selection
World
    finite resource, 6, 8
    game, 3–5, 17, 46
    infinite resource, 6, 8, 191
    perceived, 11, 111
    virtual. *See* Virtual world

Z
Zeitgeist, 16, 158

Printed in the United States
By Bookmasters